GOING
MOBILE

GOING MOBILE

How Wireless Technology Is Reshaping Our Lives

DARRELL M. WEST

BROOKINGS INSTITUTION PRESS
Washington, D.C.

The Brookings Institution is a private nonprofit organization devoted
to research, education, and publication on important issues of domestic
and foreign policy. Its principal purpose is to bring the highest quality
independent research and analysis to bear on current and emerging policy
problems. Interpretations or conclusions in Brookings publications should
be understood to be solely those of the authors.

Library of Congress Cataloging-in-Publication data is available.
ISBN 978-0-8157-2625-8 (pbk. : alk. paper)

9 8 7 6 5 4 3 2 1

Printed on acid-free paper

Typeset in Sabon

Composition by Cynthia Stock
Silver Spring, Maryland

To the inventors and entrepreneurs
who powered the mobile revolution

CONTENTS

LIST OF TABLES AND FIGURES

Tables

Figures

ACKNOWLEDGMENTS

I AM INDEBTED TO Hillary Schaub, Elizabeth Valentini, Ashley Gabrielle, Joshua Bleiberg, and Sonia Vora for the valuable research assistance they provided on this book. Elizabeth Valentini helped write portions of chapter 9 on disaster relief and public safety. Several individuals at the Brookings Institution Press deserve a special thank you. Valentina Kalk, who directs the Press, was a tremendous source of good advice about the book. Janet Walker was very helpful in supervising the book production process. None of these individuals is responsible for the interpretations presented in this work.

GOING
MOBILE

1   MOBILE TECHNOLOGY

MOBILE TECHNOLOGY IS RESHAPING society, communications, and the global economy. With cell phones, smartphones, and tablets now outnumbering desktop computers, there has been a sea change in the way people access, use, and share information. Powerful mobile devices and sophisticated digital applications enable users to build businesses, access financial and health care records, communicate with public officials, and complete online transactions. More globally, such devices and applications have helped reduce social inequality, increased participation in civic life, and increased education levels, all of which spur national economic development.[1]

This revolution in how consumers and businesses access information, and the far-reaching consequences of such uses, represents a fundamental turning point in human history. For the first time, people are able to connect with one another in a relatively inexpensive and convenient manner around the clock. In both developed and developing countries, the growth in mobile technology has been accompanied by job creation and knowledge transfer, as well as deepened social and economic connections. With the mobile industry generating $1.6 trillion in revenues, it is important to understand how mobile telephony is reshaping our world—our social connections, economic markets, and political development.[2] It is this fundamental transformation that I explore in this book.

The Rise of Mobile Technology

Mobile technology is the fastest-growing technology platform in history. According to a GSMA Wireless Intelligence report, the number of mobile subscribers around the globe has risen dramatically, from 2.3 billion in 2008 to 3.5 billion in 2014, and is expected to surpass 3.9 billion by 2017 (figure 1-1).[3] The growth in number of mobile devices is even more dramatic, for many people have more than one cell phone, smartphone, or tablet. Thus, the total number of cellular connections exceeded 7.4 billion in 2013 and is expected to reach 9.7 billion by 2017.[4] At current growth rates it will take only two and one-half years for the next billion mobile connections to be made.

High growth is especially the case in the developing world as users have skipped the desktop and laptop phases of information technology and shifted directly to handheld devices. People are using cell phones, smartphones, and tablets for communications, commerce, and trade.[5] According to Jenny Aker and Isaac Mbiti of Tufts University, mobile devices represent a significant enabler of economic development,[6] creating many opportunities for entrepreneurs and businesspeople.

The dramatic switch to mobile technology becomes evident if one examines the trend lines for installed mobile devices and personal computers, which crossed at the end of 2012.[7] As shown in figure 1-2, the total number of Internet protocol (IP) network-enabled desktops, notebooks, and netbook personal computers in the years before 2012 exceeded that of cellular phones. As more consumers and businesses adopted smartphone technology, however, those devices exceeded the number of personal computers in 2012. Smartphone installation currently is growing at about three times the rate of personal computer installation.

FIGURE 1-1. Growth in Mobile Subscribers, 2008–17

Number

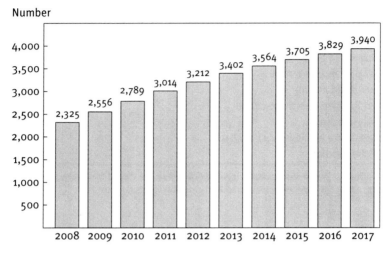

Source: A. T. Kearney, "GSMA: The Mobile Economy" (London, 2013).

FIGURE 1-2. Trends in Mobile Devices and Personal Computers, 2009–14

Number

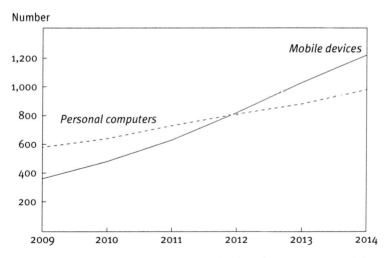

Source: Ken Hyers, "A Peek into the Future of Mobile" (Boston, Strategy Analytics, January 2011).

TABLE 1-1. Percent of Mobile Phone Owners Using Various Services, 2013

Service	Percent
Send or receive text messages	81
Access the Internet	60
Send or receive e-mail	52
Download an app	50
Get location-based directions	49
Listen to music	48
Participate in a video call or chat	21
Share location	8

Source: Pew Research Center, Internet & American Life Project Survey 2013 (Washington, 2013). The survey was conducted April 17 to May 19, 2013.

Consumers like the convenience of mobile devices. They enjoy being able to access e-mail, conduct e-commerce, and access a wide range of applications on the go. In the United States, a survey by the Pew Research Center found that 90 percent of American adults own a cell phone, 58 percent have a smartphone, and 42 percent own a tablet computer.[8] When asked whether they used various services, mobile phone users indicated that they employed their mobile devices to send or receive text messages (81 percent), access the Internet (60 percent), download apps (50 percent), get directions or location-based information (49 percent), or listen to music (48 percent) (table 1-1).

A second important trend is the growth in number of mobile broadband subscribers around the world, which surpassed that of fixed broadband subscribers at the end of 2010 (figure 1-3). It is anticipated that by 2015, there will be 3.1 billion mobile broadband subscribers worldwide, compared to 848 million fixed broadband subscribers. The extraordinary growth in mobile broadband adoption means that within a span of four years, mobile broadband will have increased to about 80 percent of all broadband subscriptions and will be the dominant means of Internet connectivity. Emerging markets have kept pace: mobile

FIGURE 1-3. Trends in Mobile and Fixed Broadband, 2010–15

Number

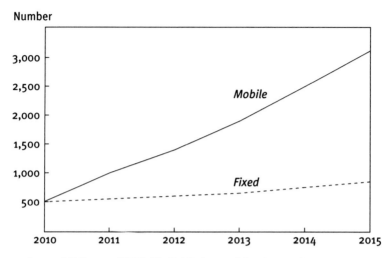

Source: A. T. Kearney, "GSMA: The Mobile Economy" (London, 2013).

broadband is expected to increase to 79 percent of all broadband subscriptions by 2015, up from its 2010 level of 37 percent.

Not surprisingly, in light of its long-term potential, a number of countries have identified broadband and wireless technology as crucial infrastructure needs for national development. Broadband is viewed as a way to stimulate economic development, enhance social connections, and promote civic engagement. National leaders understand that broadband technology is a cross-cutting technology that speeds innovation in such areas as health care, education, energy, and transportation. When combined with organizational changes, digital technology can generate powerful new efficiencies and economies of scale. Moreover, the creation of new digital platforms across a variety of domains spurs utilization and innovation and brings additional people, businesses, and services into the digital revolution. To cite only two examples, both entrepreneurs and underserved populations stand to benefit from a greater use of mobile technology.

Entrepreneurs play a major role in the economy of many countries. They launch companies, build businesses, and provide jobs. As the world moves toward a global digital economy, entrepreneurs increasingly rely on mobile technology to develop their businesses, reach markets, and pay vendors. This helps them stay in close contact with individuals and entities upstream and downstream of the business and build relationships.

Mobile technology also offers important advantages for those in underserved rural communities, where limited access to broadband and other telecommunication services makes it more difficult to participate, socially and economically, in the modern world. Mobile devices offer a way to gain Internet access even in places that are geographically remote. In larger purview, because of their relatively low cost and ubiquitous connections, mobile devices help overcome digital disparities. There are well-established inequities based on race and ethnicity, as well as certain geographic distinctions, in socioeconomic well-being. In many countries, minorities have lower levels of educational achievement and lower incomes than non-minorities. Yet mobile devices have been able to narrow the gap in technology utilization. A December 2013 report by the Pew Research Center pegged smartphone ownership in the United States at 61 percent among Hispanics and 59 percent among African Americans, both rates slightly higher than among whites (53 percent).[9] Mobile technology helps minorities start businesses, access mobile-based health care applications, engage in mobile-based learning, and otherwise tap into the benefits of the technology revolution.

The remainder of this chapter highlights the contributions of mobile technology to economic development, education, health care, civic and political engagement, and public safety. On each of these dimensions, which are taken up in turn and developed in subsequent chapters of the book, mobile technology has reshaped expectations, possibilities, and the interactions of humans both with each other and with the broader social and economic environment.

The Mobile Economy as an Engine of Job Growth and Economic Development

In the United States, mobile broadband is a significant contributor to job creation and economic development. Advanced digital infrastructure makes possible new businesses, products, and services. According to the management consulting firm Deloitte, U.S. investment in 4G (fourth-generation wireless) technology is expected to generate more than $73 billion in GDP growth between 2012 and 2016, with between 371,000 and 771,000 new jobs created.[10]

A similar picture emerges in other countries, including both developed and developing economies. For example, in a 2010 study, Jenny Aker and Isaac Mbiti found dramatic increases in mobile telephony in sub-Saharan Africa: "There are ten times as many mobile phones as landlines in sub-Saharan Africa, and 60 percent of the population has mobile phone coverage." With respect to the impact on economic development, the authors suggested "positive impacts on agricultural and labor market efficiency and welfare in certain countries."[11] Mobile phones improve welfare by reducing job search costs and making it easier for farmers and fishers to access market information.

Some specific investments have produced significant economic benefits. For example, a July 2013 study by the London-based Strategic Networks Group of the construction of a fiber-optic network in the South Dundas township of Ontario, Canada, found that an investment of $1.3 million led over several years to a "$25.22 million increase in GDP for Dundas County and [a] $7.87 million increase for the Province of Ontario," and the creation of 207 jobs.[12] The authors used a survey of businesses and organizations in the county to conclude that an increase of $3.5 million in provincial tax revenues and $4.5 million in federal tax revenues could be directly attributed to the network. Fifty-four percent of the area's businesses that had access to the fiber-optic network reported job growth, compared to 27 percent of businesses that had dial-up Internet access and 5 percent of establishments with no Internet access.

Raul L. Katz and colleagues in 2009 examined the impact of broadband technology on job growth and GDP in Germany.[13] Overall, they estimated that 968,000 new jobs and €170.9 billion would be added to the economy over ten years. This amounts to a 0.60 percentage point improvement in annual growth in the German GDP during the period 2010 to 2020 attributable to the implementation of broadband technology. Some of this growth was expected to occur in the form of jobs associated with network construction, with the remainder generated by rising broadband penetration and subsequent innovation in business activity.

A 2009 report on the GDP of 120 nations between 1980 and 2006 prepared by Christine Zhen-Wei Qiang estimated that each ten percentage point increase in broadband penetration increased the GDP of high-income countries by 1.3 percent and that of low- to middle-income countries by 1.21 percent.[14] This observation suggests that growth comes not just in direct forms but also because broadband technology spurs new applications for businesses and consumers.

Education

A number of new mobile technology initiatives are reshaping education.[15] Speaking at an education policy symposium in 2010, Mark Schneiderman, the senior director of education policy for the Software & Information Industry Association, said, "The factory model that we've used to meet the needs of the average student in a mass production way for years is no longer meeting the needs of each student," and called instead for changes in education that would recognize the magnitude of the information changes that have taken place in U.S. society. In today's world, he claimed, students "are surrounded by a personalized and engaging world outside of the school, but they're unplugging not only their technology, but their minds and their passions too often, when they enter into our schools."[16]

In the years since Schneiderman's diagnosis, the digital learning landscape has changed remarkably. Wired classrooms, hand-held devices, and electronic instructional sets let pupils learn at their own pace and in their own manner. Personalization makes education more adaptive and timely from the student's standpoint and increases the odds of pupil engagement with and mastery of important concepts. It frees teachers from routine tasks and gives them more time to serve as instructional coaches and mentors for students.[17]

Smartphones and mobile devices are being utilized for educational purposes in a variety of institutions. A 2009 analysis by Jessica Briskin and colleagues of application stores for BlackBerry, the iPhone, and the Android smartphone found that popular education-oriented downloads included My Very First App, Star Walk, Ace Flashcard, Cookie Doodle, Wheels on the Bus, and Cosmic Discoveries. Several productivity-enhancing apps for use in administration, data collection, and collaboration were also prominently available.[18]

Technology has also enriched the social environment in which learning takes place. Some teachers have developed Facebook applications (and apps on other social media platforms) for personalized learning. They post comments, get reactions from students, set up meetings, and express views about the class. Research conducted at a private liberal arts university and presented in 2010 found that students enrolled in courses set up in this manner averaged an hour per day accessing the Facebook Learning Management System. Instructors commented that students responded almost immediately to messages about the course and that pupils "engaged more in questioning through Facebook messages directed to the instructor than asking them verbally in the face-to-face classroom."[19]

Chris Dede has looked at interfaces in the United States enabled by mobile-based learning for students and found three educational advantages: allowing multiple perspectives, facilitating situated learning, and transferring knowledge from one setting to another.[20] Each of these experiences enhances the learning process and allows students to gain new knowledge or apply insights to different areas.

Handheld devices enhance student learning in other ways as well. They have been found to bridge the socioeconomic gap between the haves and the have-nots, and to expose pupils to a rich array of instructional resources. Students find the use of technology engaging and report great satisfaction with mobile learning approaches. This was particularly true of underserved populations located either in geographically remote areas or in poorer districts.[21]

While most of the studies referenced to this point concerned U.S. institutions and learners, research on the educational use of mobile devices in other countries similarly provides strong evidence of the impact of technology. A project in Taiwan, for example, compared student vocabulary mastery after reading short messaging service (SMS) English lessons versus that based on reading textbooks. The analysis showed that pupils learned more vocabulary with the former than with the latter.[22]

In China, after universities introduced mobile learning platforms in their classrooms, instructors found a sharp increase in student engagement and interaction. Instructors broadcast lectures and classroom videos to students' mobile devices. Class members could either attend the live lecture in a traditional classroom or watch via their smartphones. Teachers could use software to determine how students were engaged, what text messages were exchanged, and what pupils were learning through pop-up quizzes regarding lecture materials. In surveys and analyses conducted after the conclusion of the class, educators found that participants in the mobile learning program were more likely than nonparticipants to have posted messages regarding the course. Students reported high satisfaction with mobile learning and felt that smartphone broadcasting enhanced their educational experience. Overall, more than 1,900 messages were posted on the course forum, which instructors found to be "phenomenal" and a stark contrast to the usual reticence of Chinese students in classroom discussions.[23]

Although learning through mobile technology in countries other than the United States is in an early stage of its development, considerable research, outlined in chapter 5, "Mobile Learning," points to similar outcomes in learning and achievement. Altering the social context in which learning takes place can be a powerful accelerator of the education process.[24]

Health Care Provision

Health care delivery today is dominated by physicians, hospitals, the pharmaceutical industry, insurance companies, and government agencies. Patients seeking information or care must navigate networks of providers, order prescription drugs from pharmacies, and file claims for reimbursement from either public or private insurance plans. They have no choice but to spend hours connecting the dots and working out the best health care for themselves and their families. But health care does not have to be run this way.

Let us imagine a different system, one where, with the aid of the Internet, electronic medical records, and smartphones, the patient is in charge.[25] People monitor their own weight, blood pressure, pulse, and blood sugar levels and send the results by means of remote devices to health care providers. Medical records are stored online and accessible by patients regardless of where they are in the world. They receive personalized feedback via e-mail and reminders when they gain weight, have an uptick in cholesterol levels, miss taking a prescribed medication, or experience increased blood pressure. Social networking sites support patients through discussion forums and the collective experience of other people with similar problems. Patients take responsibility for their routine health care and rely on physicians and hospitals for more serious medical conditions.

This system is not a futuristic vision but one that is well within our grasp. It would cut costs by reducing professional responsibility for routine tasks and record keeping while also facilitating improved patient care and satisfaction. The technologies for this kind of transformation of the health care system are available now in the form of cell phones, mobile broadband, remote monitoring devices, telemedicine, videoconferencing, and the Internet.

This list of basic technologies, moreover, is augmented by the advanced features of smartphones, such as web browsing, and application development. The increasing sophistication of smartphones has given rise to a variety of new medical apps that help doctors and patients stay in touch and monitor health status and needs. For example, a mobile application allows physicians to receive test results on their mobile devices. They can examine blood pressure records over time, view an electrocardiogram, or monitor a fetal heart rate. This capability allows them to detect conditions that place mother or fetus at risk. Applications like those described make physicians more efficient and speed the delivery of health care because physicians need not be in the physical presence of a patient to evaluate the patient. If a personal conference is required, physicians can use videoconferencing to speak to patients located in another locale.

The improved patient care possible through mobile technologies has already received research support. Work by Mirela Prgomet, Andrew Georgious, and Johanna Westbrook has found that mobile handheld devices have positive impacts on hospital physician work practices and patient care. When care providers were equipped with such devices, researchers observed benefits in terms of "rapid response, error prevention, and data management and accessibility."[26] These benefits were especially profound in emergency room settings, where time is of the essence in achieving a good outcome.

More broadly, beyond the individual patient and physician, mobile devices can be used to improve global health by tracking epidemics and assisting in disaster relief efforts. A majority of sub-Saharan Africa residents, to take one example, are served by cell phones with texting capabilities. A nonprofit organization called Medic Mobile seeks to use text messaging in that part of the world to track the spread of disease and help disaster relief personnel find those in need.[27] In this way, digital technology allows people to overcome the limitations of geography in health care and to access information and deliver care remotely to individual patients or populations.

Civic and Political Engagement

Fast mobile broadband promotes civic engagement and provides new ways to follow politics and government.[28] A number of organizations around the world have developed interactive mapping software that allows citizens to chart data patterns in their neighborhood or create videos or multimedia platforms that engage people in public debates.[29]

Geographic information systems (GIS) are increasingly used for purposes of civic engagement. Interactive sites allow people to map a range of social, economic, political, demographic, and policy features onto local, state, national, or international jurisdictions by matching GIS coordinates. For example, a number of cities have mapping capacities on government websites that enable site visitors to see crime or safety data broken down by individual blocks. This allows them to chart crime statistics or transportation patterns along social, economic, or political dimensions. Such information may be used to determine government spending priorities or to direct elected officials' attention to overlooked needs.

In the noncriminal arena, Renate Steinmann, Alenka Krek, and Thomas Blaschke of the Salzburg Research Institute looked at public participatory GIS applications that focus on ways to get citizens involved in civic decisions.[30] They evaluated GIS-based sites in the United States and Europe for interactivity, usability, and visualization. Among the projects analyzed were urban design visualization, resource management mapping, river basin analysis, and landscape planning. For example, the village of Bradford, United Kingdom, has online maps that allow people to zoom in and select specific features for study. The University of Salford, near Manchester, United Kingdom, employs an "Openspace" platform with 3D capacity; users "walk through" a virtual city and are able to submit design suggestions to city planners. The website of the Landkreis Freising, a district in northwestern Bavaria, Germany, allows visitors to select development options on interactive city maps.

Public officials are increasingly using mobile communications to keep in touch with constituents. For example, G. S. Hanssen in 2008 analyzed how local politicians use digital communications to engage citizens and industry stakeholders in policymaking.[31] A national survey of municipal politicians and mayors in Norway he undertook showed that e-mail is the most important communications channel between local politicians and citizens. His study also found that mayors employ e-mail in work-related communications more than other public officials do.

H. Rojas and E. Puig-i-Abril in 2009 examined the impact of digital communication technologies on political mobilization and civic participation in Colombia.[32] Using data from a random public opinion sample of Colombia's adult urban population, they documented how broadband Internet and mobile telephony aided "expressive participation" in online protests. They undertook a survey of online information usage and found a relationship between digital information acquisition and political engagement. Those who sought information from the Internet were more politically active and expressive than those who did not use the Internet. Rojas an Puig-i-Abril conclude that in developing societies with high levels of political, economic, and social conflict, digital communications represent a valuable pathway toward democratic political engagement.

Public Safety and Emergency Preparedness

Mobile devices are especially helpful for public safety and during various types of disasters. In natural emergencies, fixed-line communications often are not available, and people must depend on mobile telephony. Landlines get destroyed and telephone and electric wires go down in major storms.

For example, during Hurricane Katrina, emergency personnel relied on smartphones and handheld devices to communicate with one another and with individuals needing assistance. Businesses were closed and office buildings were submerged under water. With people at risk from crime or lack of proper medical care, mobile communications were vital to personal well-being and to getting the community back on its feet.

Conclusion

Mobile technology, especially broadband technology, is reshaping many different aspects of social, economic, and political life. Through invention and innovation, new products and services are transforming education, health care, and governance. Patients are being empowered to take responsibility for their own health, and students have tools with which they can learn 24/7. However, it is important for countries to reap the benefits of mobile technology by investing in wireless infrastructure and promoting innovation. Creating a strong ecosystem for innovation and invention should be a top priority for leaders in every country.

2 GLOBAL ENTREPRENEURSHIP

ENTREPRENEURSHIP IS CRUCIAL FOR economic development around the world. In countries such as Nigeria, Egypt, and Indonesia, micro-entrepreneurs generate 38 percent of the GDP.[1] Analyses of time-series data demonstrate that small businesses create a disproportionate share of new jobs.[2] They generate new ideas, new business models, and new ways of selling goods and services.

The role of entrepreneurship in economic development is significantly facilitated by the convenience of wireless technology. Cellular devices help people access market information, sell products across geographic areas, reach new consumers, and use mobile payment systems. Beyond these marketplace functions, they have important socioeconomic leverage through empowering women and the disadvantaged.[3] A report from the United Nations Conference on Trade and Development found that "ICT [information and communication technology] infrastructure is an increasingly vital determinant of the overall investment climate of a country."[4] Other researchers connected to the mobile phone industry have found that each "one percent increase in mobile penetration is associated with a 0.5–0.6 rate of GDP."[5]

In this chapter, I discuss how mobile entrepreneurship improves opportunities for social and economic development. I analyze the importance of wireless technology for entrepreneurship, how mobile technology improves access to capital and market information, how it helps entrepreneurs serve broader geographic areas and reach new customers, the manner in which it empowers women and minorities, and the way mobile payments stimulate economic development. I conclude by outlining the steps we need to take to overcome current barriers to mobile-based entrepreneurship.[6]

The Importance of Wireless Technology for Entrepreneurship

Wireless technology is vital for entrepreneurship and small business development. The *Time* Mobility Poll, conducted in cooperation with the American semiconductor company Qualcomm, surveyed 4,250 wireless technology consumers in eight countries—Indonesia, India, China, South Korea, South Africa, and Brazil, with the United States and United Kingdom as representatives of the developed nations for comparison. The poll found that 93 percent believed wireless mobile technology to be very or somewhat important for entrepreneurship.[7] Indonesians (98 percent) were slightly more likely to feel that way, while Americans (87 percent) were less inclined to (figure 2-1).

Among those interviewed, 81 percent reported that mobile technology helped them search for the lowest available price for something they wanted to buy, 78 percent felt it gave them access to a larger group of potential customers, 78 percent believed it helped them follow up with people, 77 percent thought it granted access to financial services information, 74 percent believed it allowed them to find where they could sell goods for the best price, and 63 percent believed it strengthened the economy in their home country.

The poll also inquired about the importance of wireless mobile technology to different kinds of business activities. Ninety-one percent of respondents believed it would improve access to financial services for businesses and entrepreneurs. Ninety-one percent thought wireless technology would encourage home-based

FIGURE 2-1. *Time*-Qualcomm Wireless Technology Poll:
Percent Thinking Wireless Technology Is Important to Entrepreneurship, 2012

Percent

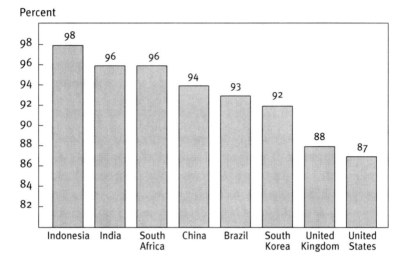

Source: "How Has Wireless Technology Changed How You Live Your Life?," *Time*, August 27, 2012, pp. 34–39. The *Time* Mobility Poll was undertaken in cooperation with Qualcomm between June 29 and July 28, 2012.

entrepreneurship by improving the ability to sell goods in regional and global marketplaces. Ninety percent believed it would help businesses use online tools such as inventory planning. Eighty-nine percent said it would enable farmers and fishers in rural areas coordinate with markets to search for the best possible price for their goods. Eighty-nine percent thought wireless technology streamlined the way businesses could handle transactions by allowing a merchant to accept payments and move funds electronically.

In nearly all developing countries, mobile phone subscriptions are growing rapidly. Countries reporting the highest growth rates between 2007 and 2012 include Armenia (115 percent), Vanuatu (113 percent), Fiji (91 percent), Botswana (88 percent), Tajikistan (82 percent), Kyrgyzstan (81 percent), and Peru (80 percent).[8] And cell phone ownership is high as well. According to a 2011 Gallup poll, 71 percent of adults in Nigeria have cell phones, as do 62 percent of adults in Botswana and 50 percent of adults in Kenya and Ghana. In Africa as a whole, there are an estimated 616 million mobile phone users, mainly because of the lack of landlines in these places.[9] The high usage levels of handheld devices create enormous opportunities for social and economic development.

Accessing Capital and Market Information

In many developing nations, it is a major challenge to gain access to capital and market information. For example, small businesses in India cite acquiring credit as their most significant problem. Few banks will lend small businesses or startups money, and many people there lack the collateral necessary to guarantee loans. As a result, entrepreneurs often have to rely on friends and family members for seed money in their businesses.[10]

Regions undergoing rapid economic development typically do not have much in the way of functioning infrastructure, and many residents lack significant financial resources. In sub-Saharan Africa, according to a study published in the *Journal of Economic Perspectives,* only "29 percent of roads are paved, barely a quarter of the population has access to electricity, and there are fewer than three landlines available per 100 people."[11] In Indonesia, 75 percent of the population have incomes below $2.50 per day.[12] The combination of poor infrastructure and poverty makes it difficult for small businesses to access financial resources and information. The lack of timely information in turn weakens pricing signals, making it more difficult for entrepreneurs or small businesses to sell their products competitively.

Individual initiatives in different developing economies have leveraged the affordances of mobile technology to encourage development and help small businesses and entrepreneurs enter and function competitively in the marketplace. In Indonesia, for example, microfranchises with mobile capability have banded together into the Ruma Entrepreneur initiative, which connects individuals with one another through shared business ventures. Entrepreneurs pool capital and secure microfinance loans to develop their businesses. Some resell airtime through mobile minutes, while others market services such as job listings or financial information. The Ruma Entrepreneur initiative has enrolled 15,000 entrepreneurs, who serve 1.5 million customers. Eighty-two percent of the businesses enrolled in Ruma are women-owned, and 47 percent claim their incomes doubled as a result of their involvement in the initiative.[13] The businesses developed included retail outlets, service companies, and clothing production, among others.

In Indonesia, farmers (mostly men) register with a mobile service called 8villages to access market information. Using a handset, they tap into a social network that gives them access to product reviews, crop prices, agricultural expertise, and weather conditions. By 2011, more than 600,000 individuals had signed up for the service and gained the benefits of crop and market prices.[14] This helped them expand their geographic scope and sell to new customers.

Services of this type have a strong track record of boosting market efficiency. A study by Jenny Aker and Isaac Mbiti of agricultural markets in Niger found that "the introduction of mobile phones reduces dispersion of grain prices across markets by 10 percent."[15] The market effects became even stronger as cell phone ownership increased. There were also significant benefits for food consumers and producers. In some businesses in that country, profits rose by as much as 29 percent annually for farmers assisted by cell phones.[16]

Farmers in Kenya, another agricultural country, have access to timely market data through an application called M-Farm. The application asks farmers to text a certain number to receive information on where they can sell their products and what the price is in different locales. This information has enabled farmers to cut out the middleman and get the best prices for their agricultural products. Around 31 million people in Kenya (around 72 percent) own cell phones, and these types of applications are very popular.[17] M-Farm's startup was funded partly through the company's winning a €10,000 grand prize in a competition for fledgling businesses, which represents a strong validation from the business community of the value of such offerings.[18]

A 2012 study by Qualcomm similarly found that mobile devices helped train fishers in Brazil and improved their access to both practical and market information. Using wireless technology, fishers gained access information on where the fish were, how to catch different species, and what water conditions were best for fishing. Financial and market applications allow fishers to both track expenses more accurately and sell directly to hotels and restaurants, thereby keeping more of the profits.[19]

Mobile technology has aided fishers in Trinidad and Tobago, where 95 percent of those in the industry use cell phones to acquire up-to-date market information. Fishers have access to two mobile applications, GFNF (Got Fish Need Fish) and Prices; the latter shows the market prices for various fishes. Seventy-eight percent of those using these applications report that the technology allows them to tap into broader market prices, as opposed to relying solely on prices offered by local dealers. Eighty percent say the applications have saved them time doing their job.[20]

The above examples help fishers move products to markets. In these and other kinds of businesses, an important aspect of business is communications, and wireless is a big help in growing markets. In Tanzania, for example, local entrepreneurs have opened Internet cafés with the aid of microfinance loans. As only 1.6 percent of the country's population have personal Internet access, these cafés give people and businesses access to the world beyond the village. People can transfer money or make long-distance phone calls, effectively opening up commerce across hundreds or thousands of kilometers.[21]

Several media companies use short messaging service (SMS) technology to provide consumers and businesses with up-to-date information. *The Standard* newspaper in Kenya has 10,000 subscribers for its business text messages, and its mobile site generates up to 300,000 hits per month. It sends four or five daily messages and charges 10 Kenyan shillings (around 11 cents) per message received. The company also markets text messages for sports and entertainment updates.[22]

The example of the South Indian fishing market demonstrates the substantial benefits achieved when mobile phones became more prevalent in Kerala. More than one million people in that Indian state worked in the fishing industry. Yet nearly all the fish were sold at a local market because fishermen had no way of knowing who the buyers were or what the prices were outside their home area. The introduction of cell phones in 1997 broadened access to market information and gave sellers up-to-date pricing signals up and down the coast. Within four years, 60 percent of the boats were relying on mobile devices to check market prices in different areas. The result, according to an article published in the *Quarterly Journal of Economics:* "Fishermen's profits increased on average by 8 percent while the consumer price declined by 4 percent and consumer surplus in sardine consumption increased by 6 percent."[23]

There are other examples as well. The Self-Employed Women's Association in India includes 1.1 million workers who pool their resources to improve their bargaining power. The organization sends agricultural workers daily SMS messages on commodity prices so that farmers can determine the best places to sell their products. Those participating say they have been able to market fruits and vegetables over wider areas and thereby earn higher incomes.[24]

The Ethiopia Commodity Exchange program has similarly helped entrepreneurs expand their markets. Before 2008, 95 percent of farmers sold their products in local markets and were not able to access other areas. Transaction costs were high, and they had problems getting fair prices because of the lack of market competition. With the advent of the exchange, though, agricultural producers gained access to external buyers and were able to negotiate better prices. This in turn boosted their incomes and improved the quality of food products by helping farmers get their items to market more quickly.[25]

Up-to-date information is similarly the focus of the services provided by the Taxi Rank, based in Capetown, South Africa. The Taxi Rank has a mobile system that allows customers to compare taxi prices for various car services. Users simply text their pickup and drop-off points, and each company sends them bids and the time at which it could provide the service. If customers have a phone with geolocation features, the company can even identify the exact pickup location.[26]

Empowering Women and Minorities

The barriers to entrepreneurship in many countries are particularly high for women and minorities, for they reach beyond market mechanics into the social fabric in which these would-be entrepreneurs live and conduct business. Nonetheless, women and minorities constitute a growing proportion of the labor force in many developing nations: in Nigeria and Indonesia, for example, women represent about one-third of the workforce.[27] Access to mobile technology is particularly important for women because globally, 300 million fewer women than men own mobile devices. Overall, the gender gap in cell phone ownership is about 21 percent worldwide, but this figure rises to 23 percent in the whole of Africa, 24 percent in the Middle East, and 37 percent in Asia.[28]

Cell phone ownership has proved crucial to women entrepreneurs as they try to break into established markets and supply chains, and thereby gain a foothold in the marketplace. It is both socially and economically challenging for them to access capital, acquire talent, and build their companies; in many countries, women do not have the same freedoms as men or the same legal opportunities to access business networks. Mobile technology has demonstrably melted these barriers to women and minorities starting and building businesses. A 2013 survey undertaken by the GSMA Development Fund and the Cherie Blair Foundation for Women, for example, found that 55 percent of women around the world "earned additional income due to owning a mobile phone" and 41 percent "increased their income and professional opportunities."[29]

Wireless communications have provided a foundation for income growth and professional development not only through their direct, practical uses in connecting businesses to customers and markets but also as they are leveraged to provide skills training in technology itself. In Indonesia, the Global Ready eTraining Center program has trained 1,000 students in technology services. Those enrolled receive vouchers for a three-month program. More than 95 percent of the individuals who have enrolled have completed the course, and 75 percent said the course increased their income as a result of the skills acquired in the program. Ten percent of the graduates have launched technology-related businesses.[30] Another example comes from India, where the Hand in Hand Partnership provides women with mentorship, training, credit, and technical support in using mobile devices to launch technology-related businesses. In Tamil Nadu, where the partnership is based, 500,000 women have joined self-help groups and started 334,000 businesses. According to the International Center for Research on Women, "monthly income has increased from 700–1500 INR [Indian rupees] to 3000–3500 INR" as a result of this partnership.[31]

Stimulating Development through Mobile-Based Payments

Mobile-based payment systems represent a way to reduce the cost of financial transactions and thereby help entrepreneurs. If people can transfer funds quickly and efficiently, it becomes easier for small and medium-size businesses to sell their products. This improves the efficiency of the marketplace and removes barriers to growth.

Reducing the friction of a financial transaction—the costs of moving money between buyer and seller, or between business and supplier—is very important in African, Asian, and Latin American financial markets because barriers to financial transactions remain quite high.[32] Only 30 percent of residents in developing African nations have bank accounts, for example.[33] A study of mobile payment systems in developing nations carried out by the Consultative Group to Assist the Poor showed that mobile-based transactions were on average 19 percent cheaper than traditional bank services.[34] The greater use of mobile payments systems, therefore, would be expected to reduce the specific barrier of transaction costs for entrepreneurs.

In Kenya, the network provider Safaricom has pioneered a successful mobile payment initiative known as M-Pesa, where M stands for mobile and "pesa" is Swahili for "money." It has 12 million people who rely on it using one of the company's 20,000 distribution sites.[35] Users can deposit money, make withdrawals, or transfer funds across accounts. It has become a popular platform for small businesses to handle their financial transactions outside the banking system. Most users find it cheaper than traditional banks and easier to transfer funds. These features in turn reduce barriers to entrepreneurship and promote business formation.

Overcoming Entrepreneurship Barriers

The various examples of entrepreneurship cited thus far have occurred, in overwhelming number, in the context of differential social and economic development. Mobile technology can help narrow the differences by reducing barriers to development. Wireless communications increase access to critical market information, improve capital access, overcome the limitations of geography, and expand market penetration. With mobile phones and tablets proliferating at a significant rate, these communication tools enable women, economic and social minorities, and other individuals to access a broader range of investors, suppliers, and customers. When these mobile devices are used in conjunction with social media platforms, people can extend their geographic reach and pool resources in meaningful ways.

Yet entrepreneurs around the world still face many obstacles. Concerns such as mobile technology costs and cultural restrictions on the use of technology are obvious impediments: in some countries, people don't have cell phones because they are embarrassed to request one.[36] Unless countries can overcome such financial, cultural, and policy obstacles, it will be impossible to realize the advantages of mobile technology.

It is also clear that mobile devices and their capabilities are not a panacea for missing elements of a standard business model. A 2012 survey of female entrepreneurs by the Cherie Blair Foundation for Women, in partnership with Exxon Mobil and Booz & Co., underscored the difference between using mobile technology and engaging a robust business model with its finding that the leading challenges for improved entrepreneurship included assistance with customer relationship management, credit, capital access, marketing, market data, financial resources, investment, training, and mentorship.[37]

This finding is seconded by a United Nations Trade and Development report on developing economies. In South Africa, for example, businesses surveyed during the study reported that their major needs included assistance with marketing (38.8 percent), gaining access to raw supplies and materials (32.6 percent), providing alternative sites (30.4 percent), providing better access to loans (29.3 percent), forming contacts with others in similar businesses (25.4 percent), easing government regulations (23 percent), gaining access to modern technology (19.7 percent), and assistance in getting loans (3.9 percent).[38]

Further, businesses function under a specific tax regime, and so it is also important to mitigate tax and investment policies that impede innovation. Because microfinance loans are a major source of capital for small businesses in many developing countries, the costs of loans and of business services should be kept low. Once entrepreneurs start to make money, the tax and regulatory policies they face should be such as to encourage business development. Keeping company registration expenses low and licensing requirements few is vital in developing economies, especially those with a high proportion of sole proprietors and micro- to small businesses.

Governments have a role to play by using their purchasing power to encourage entrepreneurship. The usual awarding of contracts to large, established firms discourages small businesses and makes it more difficult for them to succeed. Having procurement systems that are open and transparent levels the playing field and allows all to compete fairly. Electronic processes help those spread across wide geographic areas bid on equal terms for public contracts.

Legal obstacles that impede opportunities for women and minorities must also be addressed. Many individuals face limitations on starting businesses or borrowing money, and this restricts their economic development options. Legislation can encourage entrepreneurship on the part of those interested in starting companies, regardless of gender, religion, or nationality. Creating the legal structure of an incubator, where businesses can benefit from social networking and economies of scale, as well as from access to investors, has proved helpful in both developing and developed countries.

Completing this picture, and a theme touched on in this chapter, is skills training to use technology appropriately and education in entrepreneurship so that individuals learn how to develop a business model, launch the business, and continue to improve it through marketing and applied communications skills. Showing individuals how they can master the basics of business development and network with others will reduce barriers to entrepreneurial activity. If these obstacles can be overcome, many more entrepreneurs will be able to benefit from the mobile technology revolution.

3   ALLEVIATING POVERTY

POVERTY IS ONE OF the most pressing problems around the world. According to statistics from the World Bank, nearly one quarter of the global population lives at or below the poverty line of $1.25 per day.[1] With so many people struggling for basic subsistence, it is hard for those affected to lift themselves out of poverty, gain access to capital, or develop small firms or businesses that help them build a better life.

The growth of mobile technology offers options for the poor by providing new opportunities for individuals and small businesses to advance economically. Handheld devices can be used to make monetary transfers, arrange for microfinance loans, establish small enterprises, and improve the economic situation of an individual owner or family. The affordances of wireless technology thus extend to alleviating poverty and helping individuals establish themselves on a better economic footing.[2]

Jeffrey Sachs, director of Columbia University's Earth Institute, has described wireless communication devices as a breakthrough technology that contributes to solving the worst problems associated with inadequate or absent health care, poverty, and lack of access to education. In studying community life, he observed, "Now in every village where I go, someone's got a cell phone, somebody can make an emergency call, someone can find out the price on the market, someone can start a business empowered by the fact that they can reach a customer or a supplier, someone can drive a taxi or a truck for that reason as well."[3]

In this chapter I investigate how wireless technology is alleviating poverty in developing economies, particularly in its function as a facilitator of individual entrepreneurship and a driver of small business development. Despite the omnipresent barriers to doing business, such as corruption, lack of transparency, lack of capital, and poor infrastructure, in many parts of the developing world there are successful ventures enabled by mobile technology. Several examples drawn from developing economies across the globe, chiefly Asia, the Indian subcontinent, Africa, and South America, illustrate the possibilities of using wireless technology to alleviate poverty both for the individual and the country as a whole.

Opportunities for and Barriers to Advancement

The rise of mobile technology has created particular opportunities for businesses. A recent survey found that many people believe that mobile technology has enlarged their business possibilities. The 2012 *Time* Mobility Poll conducted in cooperation with Qualcomm and discussed in chapter 2 asked (among other questions) whether respondents believed that mobile technology provided access to a larger group of potential customers. As shown in figure 3-1, 84 percent of Chinese respondents surveyed thought that mobile devices had done so, as did 82 percent of Indian respondents, 73 percent of Brazilian respondents, 70 percent of South Korean respondents, 62 percent of respondents in the United Kingdom, and 53 percent of respondents in the United States.[4] These numbers underscore people's beliefs regarding the power of mobile technology to transform economic development.

Along with the opportunities associated with the use of mobile technology in economic applications, though, there are many barriers to economic progress and poverty alleviation in developing

FIGURE 3-1. *Time*-Qualcomm Wireless Technology Poll:
Percent Believing Mobile Technology Enlarges Customer Base, 2012

Percent

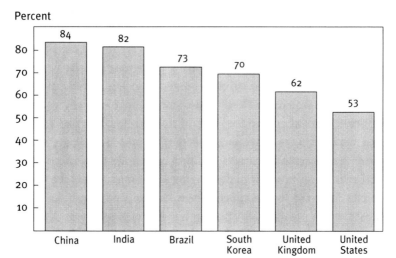

Source: "How Has Wireless Technology Changed How You Live Your Life?," *Time*, August 27, 2012, pp. 34–39. The *Time* Mobility Poll was undertaken in cooperation with Qualcomm between June 29 and July 28, 2012. The margin of error for the survey as a whole is plus or minus 1.5 percentage points.

economies. Problems such as corruption, poor infrastructure, limited transparency, and weak financial institutions impose unnecessary constraints on individuals and businesses, making it more difficult for them to take advantage of economic opportunities and suppressing economic advancement.

Corruption represents a severe problem in many parts of the world. It imposes additional costs on individual transactions and makes it difficult for those without connections to build companies. Businesses can never be sure where or when problems will arise or how demands for bribes or special payments will affect delivery schedules. An analysis by the World Economic Forum found that of 183 countries studied by Transparency International, 134 were considered "corrupt to a significant degree."[5]

A June 2010 United Nations Global Compact report found that corruption costs about 5 percent of global GDP, or around $2.6 trillion each year.[6] The following year a study by Mehmet Ugur and Nandini Dasgupta of the British EPPI-Centre at the University of London found that "a one-unit increase in the perceived corruption index is associated with 0.59 percentage-point decrease in the growth rate of per capita income."[7] The drag on the economy as a result of corruption is felt most severely at the level of the individual when extralegal payments are demanded for simple business transactions. The high cost of corruption poses an often insurmountable barrier to those living below the poverty line, who have no excess to give for a simple permit, for example.

In a number of countries, poor infrastructure is a barrier that slows economic development. Impassable roads, collapsing bridges, inadequate or contaminated water supplies, and unreliable electric grids represent a huge challenge to business formation because most developing nations lack funds for infrastructure improvements and are unable to borrow money to pay for capital improvements. The World Bank calculates that developing countries need $1.1 trillion in annual expenditures to fix their failing infrastructure. For low-income nations, this comes to 12.5 percent of their overall GDP across the developing world.[8]

Limited transparency in terms of government operations is a problem in many places. Secrecy and lack of public information creates opportunities for bribery and corruption and makes it difficult for countries to address their infrastructure and other problems. Research has shown that there is very little transparency in real estate transactions, market fundamentals, or financial performance in a significant number of countries.[9] This makes it hard to know how to do business when basic financial and political decisionmaking are obscured by murky processes.

Considerable data support the conclusion that poor transparency is a leading problem in government performance in many parts of the world. As one example, according to the latest

Transparency International Corruption Perceptions Index, "70% of countries score less than 50 out of 100." Ninety-five percent of the nations in eastern Europe and Central Asia, 90 percent of sub-Saharan African countries, and 78 percent of countries in the Middle East and North Africa have poor ratings.[10] The strong link between secrecy and corruption weakens the ability of the public sector to address basic problems such as poverty.

In addition to the lack of transparency in political, civic, and financial matters, weak financial institutions stymie development for countries as a whole. Many developing economies are struggling in the context of limited access to capital, the low capitalization of lending organizations, and a lack of physical banking branches. These barriers constrain the opportunities for economic development, and their effects are again felt strongly at the level of the individual, making it difficult for people to advance out of poverty.

Finance for the Unbanked

One of the biggest challenges for the poor is getting access to financial services. This is particularly the case in rural areas, which often lack bank branches and other types of financial institutions. Because an estimated 2.5 billion people globally do not have access to financial services such as banks or credit cards, electronic transfers provide an important way to engage in business transactions.[11]

Mobile applications are especially popular in developing economies, in conjunction with the high prevalence of handheld devices. According to a study by McKinsey & Company, high-poverty locations are more likely to have mobile devices (50 percent penetration) than to have banks (37 percent access).[12] Handheld devices reduce the cost of financial transactions and make it possible for businesspeople to offer financial services for the poor in underserved areas. Small retailers have difficulty justifying the cost of credit card readers when mobile devices can send or receive money.

This application has proved particularly useful in Asia and East Africa, where 80 percent of global mobile money transfers take place. The Indian company Eko (http://eko.co.in) offers another example of a low-cost platform that enables instant financial transactions. It works through existing retail shops, mobile networks, and banks to extend branchless banking services to ordinary people. The company partners with institutions to offer payment, cash collection, and disbursal services. Its website states that customers can walk into any Eko retail outlet and open a savings account, deposit and withdraw cash, send money to or receive money from any part of India, buy mobile talk-time, or pay for a host of services. The only equipment required for these services is a low-cost mobile phone. The service works on all handsets, and users effect transactions by dialing a number and receiving a confirmatory SMS text. The founder, Abhishek Sinha, launched the company "with a vision to promote mobile phones as a financial identity for people at the bottom of the pyramid."[13] Eko has 470 outlets across the country serving more than 70,000 customers, and the company has been awarded major grants to support its work, as well as IT implementation and social impact awards.

Another example is the Norway India Partnership Initiative's Accredited Social Health Activist (ASHA) program. The program involves the use of mobile phones to transfer money as well as identify health care facilities in the poor Sheikhpura district of Bihar, India. The activists are community link workers who connect women and newborns with hospital and postnatal care offered through the public health system. They carry "first-contact" kits for health care and promote immunization and family medicine.

Activists in the program are trained in the use of mobile payment systems, both to receive payment for their services and, if additionally certified, to facilitate care recipients' payment for medical services through mobile money transfers. The program saves time and money by providing an easy-to-use payment system that enables people to access health care and pay for vital services.[14] This type of service becomes one more way to address poverty.

Mobile Tools for Small Businesses

A number of mobile tools make it possible for small businesses to offer products that help impoverished people. For example, the mPedigree program (http://mpedigree.net) in Ghana partners with the country's telecommunications providers to offer pharmaceutical information. Developed by Bright Simons and Ashifi Gogo, the program uses basic text messaging to answer questions about counterfeit drugs. Counterfeit drugs—some innocuous, others toxic—account for half the pharmaceutical trade in Ghana, harming a million people every year.[15] Participants use their cell phones to check the unique code found on each product against a universal database. They are also able to request information on how to distinguish real from counterfeit drugs and to verify the safety of particular medicines. By 2013, the organization had placed verification codes on 10 million packs of Ghanaian medicine and expanded into four other countries.[16] The service has helped address poverty and build consumer confidence in the health care system.

Another Ghanaian service, Farmerline (www.Farmerline.org), was organized by two students, Alloysius Attah and Emmanuel Addai, who came up with the idea during a Mobile Web Ghana Entrepreneurship program. Farmerline is a mobile platform that provides "improved information access, record keeping services and communication pathways for smallholder farmers and agricultural stakeholders."[17] Farmers on this platform use cell phones to access market information and agricultural expertise provided by extension agents. The information enables food producers to enhance their productivity and sell their products to major marketing centers around the country. In addition, the company trains women in the skills needed to improve agricultural and fishing production. In Ghana, women are actively involved in the record-keeping, processing, and marketing arms of the food industry. They use Farmerline to learn about best practices and ways to develop small businesses.[18]

Etisalat (www.etisalat.ae) is an example of a useful mobile service. Operating in eighteen countries in the Middle East, Asia, and Africa, it enables users to deploy mobile phones as payment instruments. Etisalat's goal is to help businesses reach new markets and sell products around the world. Rather than having small businesses limited to local or regional markets, it has created global networks for trade and commerce.

In the United States, a San Francisco–based firm, Bizness Apps, markets mobile applications for small businesses in the developing world. It brings high-end business applications to small firms and helps them with marketing, scheduling, customer relations, social media outreach, and customizing specialized tools. The software available through Bizness Apps is cheaper than proprietary software and allows new companies to compete effectively with more established firms. So far, Bizness Apps has created over a thousand applications, and the company is active in more than twenty countries.[19]

Similarly, GoMobiMe is a Texas-based company that specializes in creating mobile-platform websites for small businesses outside the United States. Founder Joseph Stevens launched the company after learning that more than 70 percent of Internet searches were made using mobile devices and that one-third were seeking information on local organizations. He builds mobile websites for small firms in the developing world and integrates social media into his service offerings.[20] His goal is to market a product that helps people find services that improve their lives.

Microfinance

Microfinancing helps address global poverty by enabling individuals to finance small businesses and contribute to the overall economy. Firms lend small amounts of money to needy people to help them buy seeds for agriculture or to purchase equipment or raw materials. Small loan programs give low-income individuals access to capital that otherwise is hard to arrange. Through microlending, individuals are able to create companies and move themselves and their families into a more stable economic situation; they also acquire experience in managing business finances and credit.

The Microfinance Information Exchange (MIX, www.TheMix. org) operates around the world and provides microfinance performance data for developing countries. The exchange focuses in particular on the assessment of businesses that serve poor people. It seeks to strengthen financial inclusion and promote analysis of the microfinancial sector. MIX provides performance information on microfinance institutions, funders, networks, and service providers in every part of the world.

KIVA (www.kiva.org) is a nonprofit organization that provides microloans throughout the developing world. Its president, Premal Shah, launched the enterprise in 2004, along with Matt Flannery of TiVo, after taking a leave of absence from PayPal to develop the concept. The organization raises more than $1 million each week for the working poor in more than sixty countries and makes loans as small as $25. It has lent over $419 million to nearly 1.4 million people. The average loan size is about $400, and 99 percent of the loans are successfully repaid.[21] Beyond microlending, the organization has partnered with the Plasticos Foundation to provide loans for surgery for low-income people who can't afford medical treatment. Overall, that partnership has treated more than 3,000 individuals in ten different countries.

Research suggests that microfinancing programs work well, both to stabilize the economics of daily life and in the long term, to lift individuals and families out of poverty. According to the Consultative Group to Assist the Poor (CGAP), a Washington, D.C.–based consortium of thirty-four leading organizations in the finance and development fields, "Poor people with access to savings, credit, insurance, and other financial services, are more resilient and better able to cope with the everyday crises they face. Even the most rigorous econometric studies have proven that microfinance can smooth consumption levels and significantly reduce the need to sell assets to meet basic needs. With access to microinsurance, poor people can cope with sudden increased expenses associated with death, serious illness, and loss of assets."[22]

Empirical evidence from the Bangladesh Rural Advancement Committee confirms the experience of CGAP, finding that "those participating in microfinance programs who had access to financial services were able to improve their well-being both at the individual and household level much more than those who did not have access to financial services." In Bangladesh, for example, "clients increased household expenditures by 28% and assets by 112%. The incomes of people aided by the Grameen Foundation [whose efforts are focused on the poor] were 43% higher than incomes in nonprogram villages."[23]

These types of programs are especially helpful to women. In many developing economies, women have difficulty securing loans from traditional financial institutions. The combination of social norms, cultural values, and religious practices often poses an insurmountable barrier. Research by CGAP has found that "access to financial services has improved the status of women within the family and the community. Women have become more assertive and confident. In regions where women's mobility is strictly regulated, women have become more visible and are better able to negotiate the public sphere. Women own assets, including land and housing, and play a stronger role in decision making."[24]

Ways to Enable Business Development and Alleviate Poverty

This chapter has examined several ways in which mobile technology can be turned to helping individuals out of poverty, whether through education, market information, or money transfer services in areas without traditional banking systems. Under the right circumstances, mobile applications and services help producers, consumers, and markets thrive in developing economies.

Yet all of these applications and services exist within a broader economic context that frequently hinders more robust development. In many countries sweeping reforms are needed to reduce corruption, build infrastructure, improve transparency in politics and finance, empower women, and strengthen financial institutions. Wireless technology has begun to make a dent in such seemingly complex problems.

Corruption imposes a de facto tax in many nations. Corruption reduces incentives for business development and undermines public confidence in business and government. Improving transparency in the realms of politics and finance would help economies grow more quickly and move people out of poverty. Mobile websites now exist that make it possible for those asked to pay bribes to report the official's name, date, and bribe amount. Sites such as IPaidABribe.com anonymously publicize illegal solicitations and thereby bring public attention to bear on these practices. An India affiliate has received more than 400,000 reports from ordinary citizens through this site.[25]

Creating infrastructure also is necessary to improve the environment for development. As Jenny Aker and Isaac Mbiti have argued, having effective mobile networks for commerce and trade will not help unless there are functioning roads, schools, irrigation systems, and power generation facilities.[26] Entrepreneurs and businesses need these types of provisions in order to create jobs and transact basic services.

Empowering women is widely regarded as a critical step toward alleviating national poverty and improving a country's GDP, and the affordances of wireless technology are central to this effort in developing economies. Empowering the half of the population that typically faces significant barriers can unleash tremendous innovation in the developing world. Women are major beneficiaries of mobile entrepreneurship programs and microlending. Excluded from many traditional sources of capital, they often become a major source of new ideas and new energy when they find alternative ways of funding projects and launching new enterprises.

Despite the growth of mobile technology, however, many women in the developing world have not benefited from this technology. According to a report from the GSMA Development Fund and the Cherie Blair Foundation for Women titled "Women & Mobile: A Global Opportunity," women in the developing world are 21 percent less likely than men to own a mobile phone.[27] Yet in developing economies, women often manage the family's finances. As mobile usage among women rises, women will benefit enormously from mobile money services and from having greater access to financial capital. Access to capital and credit delivers the potential for lower interest rates, larger loan amounts, and opportunities for entrepreneurship and business development.

The numerous ways in which mobile technology can provide the tools, access, and services necessary to enable women's leadership and empowerment have captured the attention of the international aid community and business leaders. From mobile-phone-based mentorship of women business owners to aiding women in selling airtime minutes and application services in their communities, programs exist to accelerate women's ownership of mobile phones and provide life-changing services for women in the developing world.

In Malaysia, Qualcomm has collaborated with the Cherie Blair Foundation for Women to support women entrepreneurs. The project uses smartphones to provide training and mentoring to build ICT (information and communication technology) and English-language skills. Participants are provided with access to technology training and an electronic platform to network and share information with other like-minded entrepreneurs, as well as access to information and dedicated mentors who have the business, marketing, and technical skills to advise participants on their business goals.[28]

Women and economic minorities working to advance themselves and their families out of poverty in developing economies would benefit from stronger financial institutions. Right now, many developing countries suffer from poorly capitalized banks, few branches of national financial institutions, and limited access to capital. Under these conditions, it is very difficult for entrepreneurs to implement their ideas and build better lives for their families. Without basic financing, most small businesses will fail, and this discourages others from trying to launch new business ventures.

Company formation and registration procedures also need to be streamlined and freed of the excess registration burdens and the weight of corruption that dampen attempts to start new businesses. Making it easier to register companies and attract capital would spur business formation in many developing economies. Once proper access to capital is available and mobile applications are implemented to surmount hurdles, entrepreneurs and established businesses will find new ways to enhance global economic growth.

4 INVENTION AND THE MOBILE ECONOMY

IN 1967, INVENTOR Martin Cooper was working on a portable communications device for Motorola. The Chicago Police Department needed handheld radios, and he was using his background in electrical engineering to develop a device. As he explained in a phone interview with me, "All good inventions try to solve a social need." In the course of experimenting with mobile products, he came up with the idea of a phone that would operate over a cellular network. By 1973, he had created such as device and placed a phone call with it.[1]

This example shows the importance of invention to mobile communications. Whether it involves cellular networks, microchips, scrolling, batteries, or antennas, invention has been a big part of mobile technology. In the forty years since Cooper's invention, creative engineers have developed a range of new products that enable mobile activity.

A key tool to facilitating these advances has been the unique processes and role of voluntary standards-setting organizations (SSOs). These organizations have helped build consensus around the best ideas and corresponding technological solutions that ultimately form the basis for high-quality, interoperable networks and devices. It would be commercially impractical for mobile networks to operate without technical standards that allowed devices to function across an array of devices, multiple platforms, and numerous geographic lines. Experts from countless companies around the world have put tremendous energy, time, and resources into evaluating what ideas are most promising for global interoperability. Within these organizations, supermajorities are required before specifications are approved, published, and incorporated into products by implementers. Once certain approaches have been agreed on and adopted on a widespread scale, companies and inventors have a common basis for commercializing their products.

The global standardization of technical standards and associated work by SSOs has been crucial, as have been Internet protocol (IP) policies that ensure access to the standardized technologies. Standardization allows the industry to grow in an efficient and effective manner and has resulted in a robust, competitive marketplace. These favorable attributes of SSOs have resulted in the inclusion of more and more complex functions in products at lower and lower prices, greatly benefiting society.

As a consequence, wireless technology has become one of the most vibrant drivers of economic development. The mobile industry contributes significantly to GDP growth and job creation around the globe. In many countries, it is one of the fastest growing business areas. With mobile devices spreading at a rapid pace, it is important to understand how progress has been made and what needs to be done to facilitate future development.

In this chapter, I look at key inventors, how different countries handle invention, and barriers that need to be overcome in order to promote mobile invention. I argue that invention is critical to the future growth of mobile technology. To guarantee continued prosperity, we need to maintain the culture of invention that has propelled the mobile industry to the economic forefront. This involves making needed investment in research and development (R&D); devising better ways to commercialize knowledge;

promoting education in the STEM fields of science, technology, engineering, and math; reforming the immigration system; and maintaining a sound patent system.[2]

Key Mobile Inventors

Cellular communications are the fundamental key to mobile devices. The transmission of radio waves liberates phones from fixed landlines and makes wireless phones possible. Other crucial ingredients have come in the form of microchips, connectivity, miniature batteries, and antennas for small-scale devices. Each of these advances has promoted particular kinds of wireless innovation and made it possible to develop mobile communication devices.[3]

A number of individuals have played a crucial role in the development of mobile technology. One important inventor is Jesse Russell. He is a major figure in mobile communications because of his innovative work on cellular base stations, digital phones, and digital software radio. He was hired by AT&T Bell Laboratories and led its cellular team. Among his key patents are ones for Base Station for Mobile Radio Telecommunications Systems (1992), the Mobile Data Telephone (1993), and the Wireless Communication Base Station (1998).[4]

Russell told me in a personal interview in 2013 that he believes that "invention starts with discovery and comes from individuals who have special talent at seeing the future." In the case of early cell phones, one of his contributions was in moving mobile devices from cars to people. Previously, as he pointed out to me, cell phones were put in cars. The problem was that frequently when people placed a call, the recipient was not in a car and therefore was not in position to answer the call. Drawing on his own experiences growing up, he suggested there would be greater utility if people carried phones as opposed to keeping them in cars. That would "help people become more productive" and improve the utility of the cell phone. Once he had sold his team on the concept, he sketched out a way to make phones smaller, cheaper, and accessible to people wherever they were. This reconceptualization of the cell phone helped launch a new era of mobile communications.

Another inventor of note in this field is Arlene Harris, whose work has spurred the development of a number of wireless companies. Her latest invention is a virtual network operator for the Jitterbug phone. This is a mobile device for those who want a "large keypad, bright screen, and emergency response numbers."[5] It has big buttons and clear audio, and is thereby more user-friendly for older people and those who need such functions. Her company, GreatCall, "offers the Jitterbug phone (developed by Samsung) to those that want simple, easy and affordable cell phone service."[6]

Irwin Jacobs, the former CEO and cofounder of Qualcomm, oversaw revolutionary innovations in wireless technology that laid the groundwork of today's 3G mobile wireless standards. He was a professor of electrical engineering at MIT and later at the University of California at San Diego. Along with several colleagues, he commercialized code division multiple access (CDMA) technology. By figuring out how to share the radio spectrum, their inventions made cellular service more efficient and turned the mobile Internet into a reality.[7]

In looking toward the future, there are a number of promising inventions with the potential to have a major impact. For example, Harold Haas focuses on visible light communication that uses "a new type of light bulb that can communicate as well as illuminate—[to] access the Internet using light instead of radio waves."[8] He explained the principle in a recent TED (Technology, Entertainment and Design) talk on the website Ideas Worth Spreading. His system, known as D-Light, "uses a mathematical trick called OFDM (orthogonal frequency division multiplexing), which allows it to vary the intensity of the LED's output at a very fast rate, invisible to the human eye." In experiments to date, he has found "data [transmission] rates of up to 10 MBit/s [faster than a typical broadband connection], and 100 MBit/s by the end of this year [2011] and possibly up to 1 GB in the future." He points out that this invention makes it possible to "piggy-back existing wireless services on the back of lighting equipment."[9]

Meredith Perry is the cofounder of UBeam.[10] She is looking for a way to recharge wireless devices through ultrasonic waves. People often have laptops or smartphones that run out of power but do not have access to power cords. She came up with the idea of employing a "piezoelectric transducer to vibrate the air and produce an electrical current through movements in the crystals."[11] This is a way to recharge electrical equipment.

Maryam Rofougaran is vice president of radio engineering at Broadcom. She and her brother, Reza Rofougaran, launched a company that integrated Bluetooth and wifi in an affordable computer chip. The dual capability allows the chip to support "wireless radios for mobile backhaul, femtocells and other mobile network technologies."[12] The chip increases the battery life of mobile devices and improves overall performance.[13] Broadcom's Radio Engineering Division ships around 2 billion wireless radio chips each year.

Steve Perlman is the founder and CEO of Rearden. He has designed a "distributed input-distributed output" system, or DIDO for short.[14] It sends individually formatted data in distributed fashion to users. This allows DIDO servers to tailor service delivery in the most efficient manner. Early research finds that it is much faster than wifi, helps reduce dropped calls, and protects against signal interference.[15]

Behind each of these inventions is a creative person who came up with an innovative idea, patented the creation, and brought the discovery to market. According to Steve Perlman, however, the current U.S. system of patenting "dramatically hinders invention." It is slow, cumbersome, and overly bureaucratic. Having a

culture of invention was crucial to past innovation and is necessary for future innovation. Without understanding how to protect
and reward invention, it will be difficult to encourage the innovation that is needed for future development.

How Different Nations Handle Invention

There are many models for invention. Different nations have
quite varied approaches to facilitating scientific discovery based
on buying inventions or importing them, acquiring inventors, or
developing inventors through education.

Singapore is an example of a nation that buys invention by
offering lucrative salaries and startup packages for inventors from
the developed world. It identifies skilled talent from other countries and recruits those individuals to work in Singapore. The
country's Agency for Science, Technology, and Research has a
$2 billion National Biomedical Science Strategy that targets grant
funding toward university scientists at its Biopolis life sciences
center.[16] In recent years, Singapore has invited leading British and
American scientists to reside in the country and used their output
to stimulate local invention.

Canada, meanwhile, focuses on establishing a cadre of inventors through immigration. It uses a skills-based approach to talent development that identifies national priorities and leverages
immigration policies to address those needs.[17] Immigrants receive
points for meeting certain education and skills requirements that
move them to the front of the line for immigration. Once they
meet certain standards for vital skills in high-priority areas, they
are given visas to enter the country. This strategy enables Canada
to restock its inventor cohort and recruit top-flight talent from
around the world.

Mergers and acquisitions have been a way to enable invention
in recent years in the United States. By buying companies, it is
possible to gain access to patents and intellectual property assets.
In 2012, Google purchased Motorola Mobility for $12.5 billion

TABLE 4-1. Top Countries in Research and Development Expenditures
as a Percentage of Gross Domestic Product, 2000

Country	R&D as percent of GDP	Country	R&D as percent of GDP
Israel	4.4	Singapore	2.43
Finland	3.88	Australia	2.37
South Korea	3.74	France	2.25
Sweden	3.4	Slovenia	2.11
Japan	3.36	Belgium	1.99
Denmark	3.06	Netherlands	1.83
Switzerland	2.99	Canada	1.8
United States	2.9	Ireland	1.79
Germany	2.82	United Kingdom	1.76
Austria	2.76	China	1.7
Iceland	2.64		

Source: World Bank Patent and Intellectual Property Data Base, Washington, 2010.

and in the process acquired around 17,000 cell phone patents.[18]
In the increasingly litigious world of mobile technology, with
Apple, Samsung, and Google involved in major lawsuits over
rights to intellectual property, companies often seek to increase
their inventory of patents through mergers with other firms.

A number of countries use basic education, research, and
development to nurture domestic talent. Rather than acquiring
intellectual property or talent developed elsewhere they invest
in education in the hope of spurring invention. From the 1950s
through today, the United States and other nations have devoted
billions to higher education, K–12 education, biomedical education, and research infrastructure.

In most nations, R&D spending is funded by a variety of
sources: government agencies, business investments, universities,
or nonprofit organizations. Data from the World Bank identify
places that have the largest investments in R&D.[19] Table 4-1 lists
countries in terms of their R&D expenditures as a percentage of

GDP. That allows us to determine which places are investing the greatest proportion of their financial resources in R&D.

The place devoting the greatest financial effort to R&D is Israel, which allots 4.4 percent of GDP, followed by Finland (3.88 percent), South Korea (3.74 percent), Sweden (3.4 percent), and Japan (3.36 percent). The United States devotes around 2.9 percent of GDP to R&D. These locales place a high premium on scientific and technical knowledge. Their leaders understand the economic virtues of science and technology and the role advances in these fields play in international trade and development. As a result, they invest in R&D as a way to spur their overall economies.[20]

Technology is a big business in industrialized countries. In the United States, some 33 percent of employees work in science or technology. This figure is roughly equivalent to the 34 percent figure for the Netherlands and Germany but higher than the 28 percent figure for France and Canada.[21]

The global growth rate for high-tech products has increased by 6.5 percent in recent years, far higher than the 2.4 percent annual growth rate for other kinds of manufactured items. High tech now constitutes about 23 percent of manufacturing output in the United States, 31 percent in South Korea, and 18 percent in France and England.[22]

The productivity in this area has fueled considerable demand for those with science and engineering expertise. Thirty-eight percent of Korean students now earn degrees in science and engineering, compared to 33 percent of German students, 28 percent of French students, 27 percent of students in the United Kingdom, and 26 percent of Japanese students. The United States has fallen behind in this area. Despite great demand for persons with this kind of training, only 16 percent of graduates from U.S. universities have degrees in science and engineering, and this number is insufficient to fill all the current positions that are needed.[23]

In the United States in 1980, the private sector surpassed the federal government in terms of dollars spent on R&D. Commercial companies provide 66 percent of the $340 billion spent on R&D, compared to 28 percent from the federal government.[24] According to data from the National Science Board, the percentage of R&D spending funded by the federal government dropped from around 63 percent in the early 1960s to 28 percent in 2008, while funding from the private sector increased from 30 percent to 66 percent in the same period.[25]

A similar pattern is found in many industrialized nations. The commercial sector in most countries provides a far higher share of R&D funding than does the government. For example, the private sector in Japan provides 72 percent of funding. In Germany, commercial companies provide 66 percent, while in France, private enterprise provides 52 percent of research funding. In most industrialized countries, relatively little investment comes from the public sector.[26]

Ways to Promote Invention

Countries need people who create new ideas and bring those inventions to market. Many activities go into product commercialization, including developing ideas, forming companies, attracting capital, recruiting workers, marketing products, and launching new products. Inventors are crucial to this process because without the original idea, there is nothing to market or sell. For this reason, it is useful to look at patent filings because they tap into the idea formation that is fundamental to long-term economic development.

People patent ideas they believe are novel and innovative and that they hope will contribute to society and to the economy as a whole. Without strong, enforceable patents, neither inventors nor their financial backers could risk investing the years of effort and billions of dollars in research funding it has taken to create today's wireless technologies. The U.S. Constitution gives Congress the power to grant inventors exclusive rights to their discoveries for a limited time "to promote the progress of science and

TABLE 4-2. Top Countries in Terms of Patent Filings, 2011

Country	No. of filings	Country	No. of filings
China	526,412	United Kingdom	22,259
United States	503,582	France	16,754
Japan	342,610	Mexico	14,055
South Korea	178,924	Hong Kong	13,493
Germany	59,444	Singapore	9,794
India	42,291	Italy	9,721
Russia	41,414	North Korea	8,057
Canada	35,111	South Africa	7,245
Australia	25,526	Israel	6,886
Brazil	22,686	Malaysia	6,452

Source: World Intellectual Property Organization data compiled in World Bank Patent and Intellectual Property Data Base, Washington, 2011.

useful arts." Industries busily developing patentable intellectual property now account for 35 percent of the U.S. GDP, according to the Economics and Statistics Administration and the U.S. Patent and Trademark Office.[27]

Table 4-2 lists the nations with the largest number of patent filings. China is number one, with 526,412 applications, followed by the United States (503,582), Japan (342,610), South Korea (178,924), and Germany (59,444). Other nations that generate a relatively high number of filings include India (42,291), Russia (41,414), Canada (35,111), Australia (25,526), and Brazil (22,686). Some of these filings are initiated by individuals who live in those countries, while others come from companies that choose to file there for business reasons.

In recent years, there have been higher numbers of patent filings in China. Both Chinese inventors and companies that operate there have generated ideas and sought patents for them. Figure 4-1 shows changes in patent filings between 1961 and 2011 for China, the United States, Japan, South Korea, Germany, and India. Of these, the most dramatic improvement has come in

FIGURE 4-1. Patent Filings around the World, 1961–2011

Number of filings

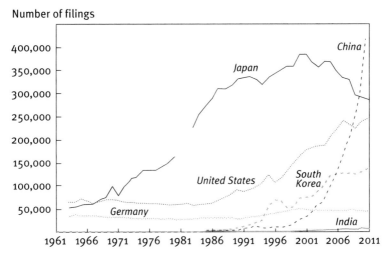

Source: World Intellectual Property Organization data compiled in World Bank Patent and Intellectual Property Data Base, Washington, 2011.

China. The number of Chinese patent applications annually went from a very few patent filings in the 1960s, 1970s, and 1980s to several hundred thousand in each of the last ten years. The United States and South Korea also have seen a rise in patent applications over this time period, while Japan has fallen and Germany has been stable over time.[28]

A number of these filings are focused on information and communications technology (ICT) and mobile innovation, which are fast-growing sectors in China and elsewhere. New applications in health care, education, communications, and economic development are being generated, and many individual inventors are filing patents to protect their ideas. These data provide a clear indication that countries focused on ICT and mobile innovation understand that securing ideas is vital to long-term development. Many countries, such as the United States and Japan, have understood that for a long time, while China now is devoting considerable attention to patent filings, having recognized the importance of invention to economic growth.

A number of factors affect the quantity and quality of invention. They include investment in R&D, the quality of education in STEM fields, the nature of immigration, and the patent system. In each of these areas, we need to maintain the culture of invention that encourages future prosperity. This involves making needed R&D investment, helping universities commercialize knowledge, promoting STEM education, reforming the immigration system, and maintaining a sound patent system.

The U.S. federal government invests around $147 billion in R&D, with $90 billion going to institutions of higher learning to underwrite faculty research projects and the training of graduate students and postdoctoral fellows. However, based on licensing fees, federal dollars generated only $2.5 billion in licensing fees for institutions of higher education in 2011. In light of the billions in government money invested in higher education research, there should be a higher yield than that for universities.[29]

Part of the problem is that universities focus more on outputs than on outcomes. Those indicators are proxy measures for getting material to the market as opposed to whether particular research ideas are actually having an impact and being successful in the marketplace. If a patent is awarded, a license issued, or a startup business established, it does not guarantee that the product is used or generates revenue.

In judging performance, most current university reporting approaches are inadequate for determining the efficiency and optimum use of research investments. There is no way through tabulations of patents and startups to measure money in versus money out with respect to university research investments. Public and private donors invest considerable funding in support of faculty work, and backers need better information to determine whether universities are making the most effective use of external resources or whether new models would produce better results. With improved metrics, it should be possible to envision alternative approaches or different personnel configurations and resource allocations.

A second problem is the insufficient number of visas granted to those who could boost innovation and national productivity. We need an immigration system that encourages entrepreneurs, allows those with graduate training in the STEM fields to remain productively in the United States, and promotes the recruitment of high-skilled talent. National leaders should elevate brains, talent, and special skills to a higher plane in order to attract individuals with the potential to enhance invention. The goal is to boost the national economy and to bring to the United States individuals with the potential to make significant contributions and increase prosperity down the road.

Right now, only around 15 percent of annual visas are set aside for employment purposes. Some of these go to seasonal agricultural workers, while a small number of H-1B visas (65,000) are reserved for those in "specialty occupations," such as scientists, engineers, and technological experts. Individuals who are admitted on an H-1B visa may work for up to six years in the United States on this form of permit, and are able to apply for a green card if their employer is willing to sponsor their application.

The number currently reserved for scientists and engineers, 65,000, is drastically below the figure allowed between 1999 and 2004, when the federal government set aside up to 195,000 visas each year for H-1B entry. The idea was that scientific innovators were so important to long-term economic development that the high number set aside for those specialty professions was warranted.

Today, most of the current allocation of 65,000 visas runs out within a few months of the start of the government's fiscal year in October. Even in the recession-plagued year of 2009, visa applications exceeded the supply within the first three months of the fiscal year. American companies were responsible for 49 percent of the H-1B visa requests in 2009, up from 43 percent in 2008.[30]

High-skill visas need to be expanded back to the 195,000 figure because at its current level, that program represents a very small percentage of the overall work permits granted each year by the United States and is woefully inadequate in terms of the supply needed. We need immigration policies that enable U.S. companies to attract top people to domestic industries and that broaden the path to invention and entrepreneurship.

Other countries, such as Canada, the United Kingdom, and Australia, view immigration more strategically, as a way to attract foreign talent. These countries explicitly target foreign workers whose skills are in short supply and who can contribute to the economy of the host nation. Their percentages are nearly the reverse of national policy in the United States, where visas for family reunification far outweigh those allocated to employment entries. Unlike other countries, whose leaders understand the value of skilled labor and occupations in short supply for long-term economic development, the United States continues to place a very low priority on admitting immigrants with special skills.

Furthermore, the United States needs to maintain a patent system that rewards invention. As Martin Cooper has said, it is important for inventors to "garner the fruits of their invention through profits." Our current system is broken and needs reform, he said. "If the patent system is improved, the invention process would be stimulated," he argued. Randall Rader, the chief judge of the U.S. Court of Appeals for the Federal Circuit, has argued that the goal is to have a system that will promote scientific progress and useful arts.

Under the system currently in place, it takes a long time for patents to be approved. The U.S. Patent Office has sought to expedite patent processing. But Congress has hindered progress by using patent fees for other purposes, according to Steve Perlman. The agency should be able to keep the fees it collects from inventors.

The Patent Office has hired new examiners and opened branch offices outside Washington, D.C. It is attempting to reduce the backlog of patents filed and has recruited reviewers with a greater understanding of patents and trademarks. Its goal is to hire more experienced intellectual property professionals.[31] But more progress needs to be made in these areas.

The patent system needs to preserve the protection of intellectual property as well as ensure the use of new ideas. Inventors need guarantees that they will benefit from their creations. We need a culture that values invention. Jesse Russell pointed out in an interview that we should "put inventors on the same pedestal as doctors, lawyers, entertainers, and athletes." It is important to inculcate these values early in children so they understand "you are doing something of value in invention." If we can do that, he argues, it is possible to maintain a culture of invention that encourages innovation and promotes long-term economic prosperity.

5 MOBILE LEARNING

EDUCATION IS AT A critical juncture in the United States. It is vital for workforce development and economic prosperity, yet is in need of serious reform. American education was designed for agrarian and industrial eras and does not provide all the skills needed for a twenty-first-century economy.[1] This disjuncture creates major problems for young people about to enter the labor force.

Mobile learning represents a way to address a number of the educational problems the United States faces. Devices such as smartphones and tablets spur innovation and help students, teachers, and parents gain access to digital content and carry out personalized assessment, vital for a postindustrial world. Mobile devices, used in conjunction with near universal 4G/3G wireless connectivity, are essential tools to improve learning for students.[2]

In this chapter, I look at ways in which mobile devices with cellular connectivity improve learning and engage students and teachers. Wireless technology is a means to provide new content and facilitate accessing information wherever a student is located. It helps learners engage with content and enables learning in ways that transform the learning environment for students inside and outside the classroom.

Unfortunately, not every student has access to a computer and the Internet. And because of the high costs of hardware, school districts are rarely able to provide a personal computer to every student. However, most young people have cell phones, and this technology provides a real opportunity to transform instruction.

As mobile phones, tablets, and other connected devices become more prevalent and affordable, the opportunities for wireless technology to dramatically improve learning and bring digital content to students greatly expand. Students enjoy mobile technology and use it regularly in their personal lives. It is no surprise that young people want to employ mobile devices to make education more meaningful and to personalize it for their particular needs.

Technology-rich activities can sustain high levels of student engagement and peer collaboration compared to less technology-focused activities. Educators need to figure out how to harness mobile platforms for instructional purposes and employ them to boost educational learning. A survey by the education non-profit Project Tomorrow found that a majority (52 percent) of students queried in grades 6–12 believed that having access to a tablet computer was an essential component of their schooling. Fifty-one percent of school administrators agreed with that sentiment as well.[3]

The United States faces the task of educating the next generation of scientists, inventors, engineers, and entrepreneurs. Educating a workforce that is effective in a global context and adaptive as new jobs and roles evolve will help support the nation's economic growth. Mobile learning makes it possible to extend education beyond the physical confines of the classroom and beyond the fixed time periods of the school day. It allows students to access content from home, communicate with teachers, and work with other people online. The value of mobile devices is that they allow students to connect, communicate, collaborate, and create using rich digital resources.

Comparison with Other Nations

The United States is at risk of falling behind other countries in educational attainment. Recent studies have found that it ranks sixteenth among developed nations in the percentage of its population aged twenty-five to thirty-four years with college degrees.[4] Korea leads the nations on this list, with 63 percent of the population in that age group holding a tertiary degree, followed by Canada (56 percent), Japan (56 percent), and Russia (55 percent) (figure 5-1). By comparison, 41 percent of Americans hold tertiary degrees. This figure is slightly above the 37 percent average for Organization for Economic Cooperation and Development (OECD) countries.

There are other signs as well. Some analysts have examined the Programme for International Student Assessment (PISA) test scores of fifteen-year-olds in sixty-five countries, organized by the OECD. PISA scores put the United States fourteenth among selected countries in educational performance. On aggregated test scores in reading, mathematics, and science literacy, American students scored 496 on the country-wide test (maximum score possible, 800 points), well below the average score of 545 for Hong Kong, 543 for Finland and Singapore, 541 for South Korea, and 529 for Japan.[5]

An analysis of international data by researchers at Harvard University concluded that "the gains within the United States have been middling, not stellar. While twenty-four countries trail the U.S. rate of improvement, another twenty-four countries appear to be improving at a faster rate. Nor is U.S. progress sufficiently rapid to allow it to catch up with the leaders of the industrialized world."[6]

The United States has, however, made some recent gains. For example, a 2013 report by Tom Loveless, senior fellow at the Brown Center on Education Policy, of Progress in International Reading Literacy Study (PIRLS) and Trends in International Math

FIGURE 5-1. Population Holding Tertiary Degrees, 2013

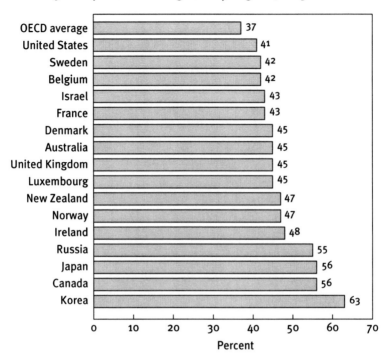

Source: Susan Lund and others, "Game Changers: Five Opportunities for US Growth and Renewal," McKinsey Global Institute report (New York: McKinsey & Co., July 2013), p. 111.

and Science Study (TIMSS) test scores showed that U.S. students had gains over the last decade in math, reading, and science, while Finnish pupils had declines in educational performance.[7] And although U.S. math scores had increased by seventeen points since 1995, U.S. fourth graders, with a score of 541, still ranked below five Asian countries in mathematics (Singapore at 606, Korea at 605, Hong Kong at 602, Chinese Taipei at 591, and Japan at 585).[8]

In an effort to make sense of these findings, the Harvard education professor Martin West has investigated the source of cross-country differences in educational performance. According to his analysis, "the evidence is clear that the American education system ranks in the middle of the pack, at best, among industrialized countries."[9]

Many reasons have been offered to explain the middling performance of U.S. students, including educational expenditures, state differentials, student diversity, parental education, and cultural factors. Martin West believes that closing the domestic achievement gap is very important and stresses the importance of improving the quality, but not necessarily the quantity, of education that students receive.[10]

Regardless of which data are studied, the bottom line is that the United States has work to do in the education area. Students in a number of other countries are performing at higher levels and generating outcomes that threaten the long-term economic prosperity in the United States. Unless the United States does a better job of innovating education, future generations will not experience the economic well-being of the current generation.

Challenges Facing U.S. Education

Education in the United States faces a number of challenges, among them the need for better digital infrastructure, more personalized content, embedded assessment possibilities, and the professional development of teachers.

Better Digital Infrastructure

Most people think of infrastructure in terms of highways, bridges, and dams. Historically, those items have consumed the bulk of U.S. government infrastructure spending. For example, the United States spent $425 billion to construct the 47,000 miles of the interstate highway system in the 1950s and 1960s.[11]

In a digital world, access to universal, high-speed mobile networks is crucial for economic development, and such networks should now be considered part of infrastructure. Educational institutions need high-speed electronic platforms, as do libraries, hospitals, businesses, and nonprofit organizations. Fixed connectivity is inadequate to the needs of today's schools, libraries, and hospitals. The connectivity must be available wherever students and teachers happen to be. Cellular connectivity, with its near universal coverage, can enable learning and promote education.

To make connectivity useful for education applications, it should operate at high download speeds and have broadband functionality. According to the 2010 Federal Communications Commission (FCC) National Broadband Plan, the United States should boost download speeds to at least 100 megabits per second for businesses and households by 2020 and 1 gigabit per second for anchor institutions in local communities, such as schools, hospitals, and libraries.[12] Such a network will not be cheap to put in place. U.S. government officials estimate it would cost around $350 billion to develop a universal high-speed Internet network.[13]

The focus of much of the National Broadband Plan is fixed access, but in a chapter devoted to education, the plan also recognizes the importance of cellular access. One of the plan's recommendations is "The FCC should initiate a rulemaking to fund wireless connectivity to portable learning devices. Students and educators should be allowed to take these devices off campus so they can continue learning outside school hours."[14]

According to Project Tomorrow data, 68 percent of high school students surveyed in 2013 said they accessed the Internet via a 3G/4G mobile device—including a number of students who did not have broadband access at home.[15] As Julie Evans of Project Tomorrow noted, "This is the real story as to why mobile devices can help to solve the home broadband problem." She went on, "Even amongst students who say they have high-speed Internet access at home, in many cases though students never get to use that access. If there is one family computer that is hard-wired for that high-speed access, students today need to contend for access with siblings who are also trying to do their homework, parents looking for jobs or doing their own work, and family entertainment activities using that computer. Students tell us that having their own mobile device that is not a shared device gives them better, more reliable access to the Internet than trying to use the family broadband connection."

Many public schools typically do not have sufficient IT staffing and so are unable to take advantage of recent advances in mobile networks. Many rely on students and teachers to bring their own devices. This situation complicates the education IT environment because schools end up with a myriad of devices and operating systems, which makes it difficult to link hardware and connect students to each other. The result is an IT tower of Babel, with multiple interoperability problems and poor communication. Above all, "bring your own device" is fine for students who have devices. Those who do not are left behind.

Personalized Digital Content

A current reality facing educators is diversity in the student body. Pupils come from different backgrounds, have different levels of preparedness, have divergent interests, and learn in unique ways. Mobile devices and technology hold promise for personalizing education within the diverse classroom and helping student cohorts advance through the system.

In their individual lives, young people are accustomed to personalized content and instantaneous communication. They seek information around the clock and pursue knowledge that is relevant to their particular interests. Indeed, one of the virtues of mobile devices is their ability to provide personalized digital content 24/7.

We need to think of education as an individualized and year-round activity, not just something that takes place in mass-produced form within a school building between 8:30 a.m. and 3 p.m. Content should be both broadly available and customized so that students can follow their interests and figure out where to get answers to basic questions.

For example, mobile technology provides students with access to always-on, anytime, and anywhere learning via smartphones and tablets, with additional benefits that include increased functionality, sophisticated graphics, and a larger screen, allowing students to work with their learning materials in new and compelling ways.

Some studies show that students are more open to using technology for learning and that they are aware of new learning tools such as text, illustration, and audio and visual recordings, all of which provide children with a more holistic learning experience. While widespread improvements in technology and communications have modernized some aspects of the education system, mobile devices have the capacity to multiply technology's impact by joining massive amounts of information with a student's imagination.

Too often, there is inadequate attention to tailoring educational content to the individual needs and learning styles of students. According to an NMC (New Media Consortium) Horizon Report, "the demand for personalized learning is not adequately supported by current technology or practices."[16] Students need ways to tailor the constant flood of contemporary information to their own needs. Software can help them separate the noise from the signal and focus on material that is most relevant to them. In that way, mobile technology can help organize the flood of information that students and teachers find online.

Embedded Digital Assessment

A benefit of the digital learning applications is the ability to embed assessment tools within learning episodes. Mobile devices make it possible to compile detailed metrics on how students approach subjects, how they acquire knowledge, and how quickly they pick up on key concepts and skills. When content is delivered in digital form, it is easy to add pop-up quizzes that evaluate comprehension and knowledge.

These kinds of tools free teachers from the mundane tasks of grading rote test items and provide immediate feedback for students and parents. Software offers nuanced measures of student attainment. Teachers can develop dashboards that track classroom activities and individual student achievement on the student's learning curve. Students can be categorized into different groups based on whether they have fallen behind their peers and need remedial attention, are on par with their fellow students, or have mastered the material and need more challenging assignments to advance their learning.

Right now, pupils who need extra attention to master key concepts and major skills often don't get it, and students who learn quickly and have mastered the material are bored because they have to wait for the rest of the class to catch up. Teachers have to focus on the classroom average because that is where many of the students are in the learning process. Neither advanced students nor those needing extra help are well served under the current status quo. Having a computerized assessment embedded in the curriculum would help students at both ends of the distribution receive the help they need.

Teachers' Professional Development

Teachers are crucial to the success of any school reform or learning innovation. As the agents of learning in the classroom, they guide education in ways that are critically important. If teachers are provided with the training and funding necessary to employ helpful tools, they will be more effective in transmitting knowledge and skills to their students.

The same is true for principals and superintendents. Schools with supportive administrators are more likely to use education technology and to incorporate mobile devices into classroom learning exercises. Teachers take cues from their supervisors, and this affects their use of mobile technology and the impact such technology can have on student performance.

Too often, new technologies are introduced into the classroom with little to guide teachers as to their use and possible benefits. Teachers are expected to learn how to use new tools without any assistance or training. It takes time to figure out new technology and knowledge to implement it in ways that advance learning. An inadequate instructional level may result in stagnant educational results.

The alternative way to think about education innovation is the professional development of teachers. Like other professionals, teachers need training opportunities that teach best practices and ways to take full advantage of educational innovations. They benefit from instruction in how to use mobile technology and how it can make their lives easier while also engaging students. They may also engage in peer learning as they discuss the affordances of education-based technology with other professionals in the field.

Digital technology also helps teachers think about different classroom models from the ones they are accustomed to. As students can take more responsibility for their own learning, teachers are able to focus on more advanced problem solving and on building critical skills for those in their classrooms. The result is an educational collaboration that is more satisfying for both students and teachers.

How Mobile Technology Facilitates Innovation in Education

There are a number of ways in which mobile technology contributes to innovation in education. Some fairly obvious ways include making collaboration easier, providing more differentiated instruction, and helping students pursue individualized projects.

One of the virtues of mobile technology is how well it facilitates social collaboration. According to a study by the High School Survey of Student Engagement, "the most engaging forms of class work involve collaborative or creative components." Its authors noted that "61 percent of students report being excited and engaged by discussion and debate in class."[17] Wireless devices represent some of the easiest ways to get students to work together.

This result is supported by the National Commission on Teaching and America's Future (NCTAF), which undertook a study of learning through teamwork in the STEM fields of science, technology, engineering, and math. It found that "STEM learning teams have positive effects on STEM teachers and their teaching."[18] Students who were part of a team were more likely than others to use mobile applications in their courses and to make greater use of digital resources.

On the teaching side, in partnership with many universities across the United States, NCTAF is looking at how mobile applications help with teacher training and support, especially during the first few years of teaching, when teachers need it the most. NCTAF's Teachers Learning in Networked Communities 2.0 (TLINC) initiative uses mobile technology to strengthen support networks and decrease the isolation of new teachers. Having tools that differentiate classroom instruction facilitates communication, collaboration, and access to digital resources for teachers.

The American Association for the Advancement of Science has launched an Active Explorer application that takes advantage of mobile devices to increase student interest in learning, particularly in the sciences. In it, smartphones are used as data collection instruments in field projects. Teachers create "quests" that provide a way for students to collect data from field experiments or to document their findings for projects. Students in several different Washington, D.C., schools took part in this project and reported positive results. They felt more engaged in the research and developed a better understanding of data collection techniques.[19]

To take advantage of these opportunities for mobile learning, students need both devices and mobile access. Not all teenagers or adults own a cell phone because of restricted family budgets, though ownership has risen, especially among students. A 2013 survey by the education nonprofit Project Tomorrow found that

FIGURE 5-2. Student Personal Access to Mobile Devices, 2013

Percent

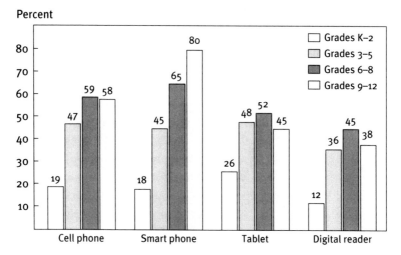

Source: Project Tomorrow, "2013 Trends in Online Learning Virtual, Blended and Flipped Classrooms," Tomorrow.org, 2013.

80 percent of high school students responding to the questionnaire said they had a smartphone, 45 percent had tablets, 38 percent had digital readers, and 58 percent had a cell phone (figure 5-2).[20] These percentages are up from just a few years earlier: in 2008, only 28 percent of high schoolers surveyed had a smartphone. And in 2011, only 26 percent of students surveyed in grades 6–8 had a tablet computer, compared to 52 percent in 2013.

Student and Teacher Engagement

Both teachers and students report positive impacts on learning through using digital technologies. For example, in the 2013 Project Tomorrow survey, teachers reported seeing a number of positive benefits from the use of digital instructional tools: 52 percent said their students were more motivated to learn, 36 percent believed their students were developing creativity, 29 percent thought the tools encouraged problem solving and critical thinking, 7 percent said their students were applying knowledge to practical problems, and 26 percent reported that students were taking ownership of their own learning.[21]

Students also express positive views on how they think the use of mobile devices will transform their learning environment. According to a Project Tomorrow Speak Up survey of middle school students, using a mobile device in school helps them increase their learning: 78 percent of respondents in grades 7 through 9 said it allowed them to check grades, 69 percent credited it with helping them take class notes, 64 percent enjoyed using mobile devices to access online textbooks, 56 percent said it helped them write papers and do homework, 56 percent used it for calendar updates, and 47 percent indicated it helped them learn about school activities.[22]

Relying on a mobile device in school, furthermore, helps students "leverage the device's capabilities to increase the personalization of the learning process." With it, 73 percent of middle school pupils said it helped them undertake research "anytime, anywhere," 63 percent praised the fact that they could receive reminders and alerts, 61 percent believed it aided collaboration with peers and teachers, 54 percent credited it with helping them organize their schoolwork assignments, and 52 percent said mobile devices allowed them to access their school network from home.[23]

To aid schools in the use of mobile technology and the educational techniques it facilitates, companies such as kajeet, a wireless service platform solution provider, are working to provide access to applications and tools to further enhance students' learning experience through mobile technology. The Making Learning Mobile pilot program is aiding 120 fifth graders at an underserved elementary school in Chicago Public Schools with the latest Samsung Android tablets to provide round-the-clock access to educational resources.

These devices utilize the company's web-based Sentinel platform, which manages Internet access, allows students to collaborate on schoolwork from any location, and provides teachers with the ability to deliver one-to-one lessons using student-safe and regulatory-compliant instructional resources. As most of the students in the pilot are English-language learners, this access provides an unprecedented opportunity for the school and the teachers to extend the learning process beyond the classroom and involve parents in supporting their child's acquisition of literacy skills.

The Consortium for School Networking (CoSN) has launched an initiative aimed specifically at providing resources and guidance to school IT and technology administrators who are working to implement wireless infrastructure and devices into their district classrooms and offices.[24] But funding remains a key barrier to the realization of these objectives. According to Keith Krueger, the CEO of the consortium, "most school districts, even those that are affluent, would say they can't afford the ongoing cost of a ubiquitous technology environment."[25]

Some schools have moved forward with their own initiatives, and the results have been powerful. One notable example is in Onslow County, North Carolina, with a project known as Project K-Nect. Since 2006, the Onslow County School District has experimented with the use of smartphones and tablets in high school math classes as a way to engage students and produce better achievement. The results showed higher scores on standardized tests in courses where students used mobile devices. Participants reported that they felt better prepared for their final exams and had increased interest in college and in careers requiring math skills.[26]

The Need for Action

Two critical factors enable mobile learning to reach its full potential—the universal availability of mobile devices and cellular connectivity for these devices. Each gives students and teachers access to a rich array of digital resources and allows them to access materials 24/7. When progress is made on the four main spheres discussed in this chapter—building digital infrastructure, expanding digital content, embedding digital assessment, and providing professional development for teachers—students are in a strong position to advance their skill levels and learn new material.

In the twenty-first-century economy, students need a range of skills beyond traditional math, reading, and writing. As Esther Wojcicki, an English teacher at Palo Alto High School, noted at a recent conference, an innovation-based economy requires people to be self-directed learners, work independently, apply technology effectively, create media products, be adaptable to change, and be good digital citizens.[27]

Others have explained how mobile technology improves the opportunities for learning. The Harvard University education professor Chris Dede has noted: "Teaching is like an orchestra. There are many different instruments, and to reach everyone you need to put a symphony of different kinds of pedagogy together. Learning technologies provide a set of instruments teachers can use to achieve that range of instructional strategies."[28]

Improved mobile infrastructure is crucial for achieving the full benefits of the technology revolution. We need better capacity in wireless networks so that students and teachers can access content and take advantage of multimedia resources that are coming online. This involves simplifying cell tower construction rules and building faster mobile networks. CIOs and technology officers in the field of education need to think about their future wireless networks. Wifi alone can't support the bandwidth requirements of heavy multimedia. Instead, high-speed wireless area networks (WANs) will be needed to take advantage of the educational resources that are available.

One thing the federal government should do is reform its E-rate program, which provides funding for schools and libraries to upgrade their Internet networks and connectivity speeds. The program was founded in 1998 to improve telecommunications access. Currently it does not include funding for WANs or mobile devices even though there are well-demonstrated benefits in terms of improving devices, connectivity, and infrastructure for educational purposes. As noted previously, the National Broadband Plan recommends that the FCC provide funding through the E-rate program for mobile devices and connectivity so that learning can occur off school grounds.

This is an important step that will help students and teachers take advantage of the opportunities for mobile learning. President Barack Obama has outlined a national goal "of connecting 99 percent of school students to the Internet through high-speed broadband and high-speed wireless within five years." According to the FCC, "more than half of the schools and libraries reported that their Internet connections were too slow to meet their needs."[29]

Some schools are using bring-your-own-device policies to overcome budgetary limitations. While requiring students to come to school with a mobile device is a quick way to boost Internet access, the policy may make no provision for individuals whose home finances do not include such devices, or who must share with siblings. The United States already has a digital divide based on income levels, and we must work to ensure that all students, regardless of economic background, have access to mobile devices. If some individuals lack access to mobile hardware and software, it robs them of the educational opportunities that are available to their classmates.

Classroom teachers report that they face a number of obstacles in using tablets in schools. They include Internet connectivity problems, slow speeds, tablet management issues, Java code challenges, finding free applications that work on all devices, and the lack of keyboards and peripherals.[30] We need better infrastructure, hardware, and training so that schools can overcome these problems.

Teacher development is at the core of educational innovation. Julie Evans, the chief executive officer of Project Tomorrow, notes that "implementation success depends upon the teacher." We must help teachers learn how to use technology to improve productivity and student outcomes. They need training in the ways technology enhances educational attainment and teacher performance. Unless they believe technology improves instruction, teachers are not likely to adopt the new approach or to implement it in ways that will be effective.

A recent report from McKinsey & Company noted that recruiting high-quality teachers was one of the most important things the U.S. school system could do to improve educational outcomes. In looking at outcomes from a number of different countries, it found that high-performing nations recruited the bulk of their teachers from the top academic talent of their country. In contrast, only 23 percent of new teachers in the United States come from the top third of their cohort.[31]

To improve this situation, the report calls for the United States to provide competitive compensation (especially in the vital fields of math and science), increase the prestige of teachers, pay for teacher training, adopt performance pay for teachers, and provide opportunities for career advancement. By elevating the profession of teaching and giving more attention to teacher development, we can increase the odds of bettering educational outcomes.

Improving the nation's digital infrastructure, individuals' access to devices, the content available for learning, and teacher engagement is important as more and more countries turn to investing in technology to help bring their classrooms into the twenty-first century.

There is already a wide range of new digital content that is available to students and teachers. This content includes instructional games, augmented reality games, interactive websites, and personalized instruction. The virtue of electronic information is that it gives students greater control over their curriculum, thereby allowing students to proceed at their own pace and in their own learning styles.[32]

The digital revolution also makes possible the real-time assessment of student performance. No longer do students have to wait weeks to receive feedback on their skill mastery. Teachers can now embed pop-up quizzes in online content delivery, and pupils can be evaluated on an ongoing basis. This strategy provides real-time feedback to students and parents and allows teachers to see which students need extra help and which ones need more challenging assignments.

Challenges to building an appropriate digital infrastructure for learning exist, but the results from a variety of pilot projects and research studies all indicate that students, teachers, administrators, and parents are excited about the opportunities and interested in exploring how to effectively leverage mobile connectivity to increase student achievement, teacher productivity, and home support of learning.

As a nation, we are educating the next generation of workers. Students in the twenty-first century need an education that includes higher-order thinking skills, analysis, synthesis, teamwork, and collaboration. The advances seen, and to come, in technology, as well as changes in organizational infrastructure, underscore the need for workers to be able to think creatively, solve problems, and make decisions as a team.

Embracing and institutionalizing mobile technology can transform learning. It is a catalyst for creating impactful change in the current system and crucial to student development in the areas of critical thinking and collaborative learning. Those are the skills that young people need to secure their place in the globally competitive economy.

6 HEALTH CARE

THE EXPANSION OF MOBILE technology offers considerable potential to improve health care. The utilization of smartphones and tablets opens up communications, information access, and service delivery. Their emergence empowers patients, health care providers, and entrepreneurs and offers new options to make transactions and access expertise. In a number of different ways, this technology is poised to alter how health care is delivered, the quality of the patient experience, and the cost of health care.[1]

In this chapter, I focus on innovations in the use of mobile devices in the practice of medicine and public health, or m-health, around the world. I review the adoption of novel m-health applications, the impact of mobile technology on service delivery and medical treatment, and the way the use of mobile devices is slowly changing the health care system.[2] I show that mobile technology helps with chronic disease management, empowers expectant mothers and the elderly, reminds people to take their regular medication at the proper time, extends service to underserved areas, and improves health outcomes and medical system efficiency.

The Rise of m-Health Initiatives

There has been an explosion of m-health activities around the world. A global survey of 114 nations undertaken by the World

Health Organization in 2011 found that m-health initiatives had been established in many countries, though there was variation in adoption levels.[3] The most common activity was the creation of health call centers, which respond to patient inquiries. This was followed by using short messaging service (SMS) texts to send appointment reminders, convey telemedicine advice, access patient records, measure treatment compliance, raise health awareness, and monitor patients, and to provide physician decision support.

Not surprisingly, there were big differentials between developed and developing nations. Africa had the lowest rate of m-health adoption while North America, South America, and Southeast Asia showed the highest adoption levels. A number of countries had initiatives in the pilot stage during the period of the study or had informal activities under way.

Member states reported their biggest m-health obstacles were competing priorities, budgetary restrictions, and staff shortages. Concerns over privacy and data security also were cited as barriers to effective implementation. To address the financial burden imposed on the government, most countries today are implementing m-health through various types of public-private partnerships.

Managing Chronic Disease and Substance Abuse

Chronic disease management and the treatment of substance abuse represent the greatest health care challenges in many places. Remote monitoring devices enable patients with serious problems to record their own health measurements and send them electronically to physicians or specialists. This practice keeps them out of doctors' offices for routine care, and thereby helps reduce health care costs.

A Brookings Institution analysis undertaken by the economist Robert Litan in 2008 found that remote monitoring technologies could save as much as $197 billion over the next twenty-five years in the United States.[4] He found that cost savings were especially prevalent in the chronic disease areas of congestive heart failure, pulmonary disease, diabetes, and skin ulcers. With round-the-clock monitoring and electronic data transmission to caregivers, remote devices speed up the treatment of patients needing medical intervention. Rather than waiting for a patient to discover there is a problem, monitors identify deteriorating conditions in real time and alert physicians.

Real-time management is especially important in the case of chronic diseases. In diabetes, for example, it is crucial that patients monitor their blood glucose levels and adjust their insulin intake to proper levels. In the days of a face-to-face delivery model of medicine, patients had to visit a doctor's lab or medical office, take a test, and wait to be informed of the results. The process was expensive, time-consuming, and inconvenient for all involved. Having to undergo regular testing for chronic conditions is one of the factors that drive up medical costs.

However, it is possible to use remote monitoring devices at home that record glucose levels (for example) instantaneously and electronically send them to the appropriate health care provider. Some patients use GlucoPhones, which monitor and transmit blood glucose levels to caregivers while also reminding patients when they need to perform the glucose test. This technology puts people in charge of their own test taking and glucose monitoring and keeps them out of the doctor's office until they need more detailed care. Health authorities estimate that more than 11 million of the roughly 24 million diabetics in the United States use home monitors to check glucose levels. This is an important advance, as diabetes is the seventh leading cause of death in the United States.

In Mexico, diabetes is the leading chronic health problem. Community health care workers employ mobile devices to answer questions and provide access to up-to-date medical information. They also point patients to mobile applications that track glucose levels, analyze symptoms, and make recommendations for care and treatment. Health providers there have had good results in pilot projects, particularly in the form of faster diagnosis and treatment.[5]

Another example of remote monitoring in relation to chronic diseases comes from Asia. China Telecom has deployed an electrocardiogram (ECG)–sensing handset that records thirty seconds of heart data and transmits that information electronically to the twenty-four-hour Life Care Networks Call Center in Beijing. That facility has thirty physicians, and the wireless monitors enable them to provide a diagnosis remotely to patients in underserved areas and to give real-time feedback to those with chronic heart disease. Cardiovascular disease kills three million individuals annually in China, and those outside the major cities have difficulty getting access to medical care.[6] The impact this program has had on improving health care earned it a Computerworld Honors Laureate award for 2012.

Substance abuse is another prominent public health area that has benefited from m-health initiatives. Researchers at the University of Massachusetts Medical School have developed remote monitoring devices called i-Heal for substance abusers. Individuals wear sensors that monitor skin temperature and nervous system activities associated with drug cravings. The devices transmit the monitoring data along with self-reported stress levels to health providers trained to look for risky behavioral and physiological patterns and to provide various types of text, video, and audio interventions designed to discourage drug use. Physicians report positive responses from participants in residential treatment programs who have used the system.[7]

A study of a smoking cessation program that incorporated applications for mobile devices found some positive benefits. Researchers identified forty-seven different applications designed to stop smoking. These apps counted the number of cigarettes smoked, suggested visualizations designed to encourage people to quit smoking, or employed quota systems or calculators that sought to reduce smoking levels. The project found "some promise" in text messaging interventions in smoking, but also discovered that many apps needed to connect to outside resources, such as health clinics or counseling services. The authors suggested that app developers adhere more closely to public health guidelines on smoking cessation.[8]

Automated Reminders

Automated reminders to take medication or to show up for an appointment are simple ways to improve disease management. In South Africa, for example, a physician was concerned that his patients did not always take the prescribed Rifafol medicine for their tuberculosis. For the treatment to be effective, people had to take the pill on a consistent basis. So the doctor set up a text messaging service called On-Cue Compliance for each of his patients that sent them a daily SMS in English, Afrikaans, or Xhosa. Over the six-month course of treatment, his service would send a message at a predetermined time each day reminding patients to take Rifafol.

In the United States, Dynamed Solutions provides HealtheTrax, commercial software that reminds patients to take medications, set up appointments, and track their own compliance with medical instructions. This and other types of "virtual health assistants" are particularly helpful to those with chronic illnesses who need to keep close track of their medical condition and stay in touch with their caregivers. The software is integrated with the patient's electronic medical record and can store information in the patient's personal records.

Dr. Robert Schwarzberg of Sensei Corporation has developed a mobile "coaching app" for persons with common chronic conditions. Using a subscription model, it offers "weight loss advice [and] virtual coaching on issues, including diabetes and blood pressure-control."[9] It tracks physical activity and body mass numbers. Based on the individual subscriber's goals, it suggests food choices based on desired caloric intake and lessons on lifestyle and weight loss.

Researchers in China found that "text message and telephone reminders improved appointment attendance by 7 percent."[10] And in Malaysia, nonattendance dropped by 40 percent for those receiving texted reminders of their medical appointments.

Simple notifications such as knowing how long they must wait before being seen by a health care professional help patients stay on track with their health care regimens. Portsmouth Regional Hospital in New Hampshire has a service designed to shorten patient waiting time for the emergency room. Patients can text "ER" to a designated number and get the anticipated wait time to see a doctor or nurse. This reduces waiting time and provides people with the time estimates they need.[11]

The overall results of using wireless-enabled technology to monitor health status and facilitate timely intervention are quite good. Error avoidance rate is improved, and health providers are relieved of some part of the burden of caring for chronically ill persons if their care is better managed. Still in an early stage are studies linking m-health initiatives to quantifiable outcome results.

For practitioners—nurses, doctors, and community health workers—error avoidance and improved service delivery are important metrics. A review of evaluation studies of mobile devices and physician practices in reading ECGs or computed tomography (CT) scans found three benefits: (1) physicians with access to mobile devices responded more promptly to reading medical test results, (2) there were fewer errors in medication prescription and hospital discharging, and (3) doctors showed improved data management and record-keeping practices.[12]

Indeed, error avoidance is one of the primary strengths of m-health. A study of nurses who used handheld devices found that 16 percent said the mobile equipment had helped them avoid at least one error in clinical treatment, while another 6 percent indicated it had enabled them to avoid errors on multiple occasions.[13]

These positive findings demonstrate that m-health has the ability to improve service delivery and save money on health care.

For patients, wireless-enabled technology improves care management metrics and promotes self-care activities such as exercise, weight management, and anxiety reduction. A metareview of studies on voice and text messaging interventions in care management looked at the results of medical reminders sent to 38,060 individuals. The researchers found improvements "in compliance with medicine taking, asthma symptoms, stress levels, smoking quit races, and self-efficacy. Process improvements were reported in lower failed appointments, quicker diagnosis and treatment, and improving teaching and training."[14]

There are also links to improved health outcomes through a combination of remote sensors and physical activity reminders. A study of a physical activity program in the United Kingdom found that real-time mobile feedback combined with wrist-worn accelerometers monitoring physical activity yielded positive results. Those getting virtual interventions showed an increase of two hours and eighteen minutes per week in physical activity and lost 2.18 percent more body fat than a control group not getting such interventions.[15]

The wireless industry has taken a leading role in developing and refining medical and care applications for mobile devices, especially Apple's popular iPad tablet. Among the most widely used apps for the tablet are ones by Airstrip Cardiology that enable physicians to view ECGs; an app by 3D4Medical called Skeletal System, which shows the human skeletal system; Orca Health's EyeDecide MD, which has optometric information; and MIM Software's Mobile MIM, which displays various types of medical images.[16] It is estimated that more than 40,000 mobile health applications exist across multiple platforms and that 247 million people have downloaded a health care app.[17] This clearly is a growing market and a sign of consumer interest in m-health.

Helping Expectant Mothers and the Elderly

Mobile technology has a large potential to improve health care provision among demographics at either end of the life span, newborns (and their mothers) and the elderly, and so to improve both quality-of-life metrics and life outcomes. The long-term results are a healthier population and a reduced burden on the national economy to care for the very ill young and old.

Wireless-enabled communications are a boon to people undergoing transient yet significant bodily changes, such as a pregnancy, as patients no longer need to visit a doctor or a nurse in person to receive care reminders or basic information on their health status. They can get personalized information by e-mail, automated phone calls, or text messages. Text4Baby is a U.S. mobile application for pregnant women. It sends text messages in English and Spanish on what changes to expect as the pregnancy progresses and how to handle different problems that may come up. In the United States, 281,000 new mothers had signed up for the service as of 2012.[18]

In Bangladesh, a developing economy, maternal mortality is a critical public health problem, and neonatal deaths account for more than half (57 percent) of deaths in persons under the age of five; fully 90 percent of childbirths in rural areas occur outside hospitals or health care clinics. To improve maternal-child survival and develop maternal awareness of preventable deaths, Bangladeshi doctors launched a Mobiles4Health initiative that provides information on pregnancy care, early warning signs of fetal and neonatal problems, the benefits of family planning, and breastfeeding best practices. Early indications suggest that such information helps expectant mothers avoid medical problems.[19]

Other initiatives have generated positive results. Several countries have launched a mobile birth notification system that calls health clinics when labor starts. Through this service, families are able to secure the services of a midwife, with the result that 89 percent of births now take place with trained health workers in attendance.[20]

Researchers at Johns Hopkins University in the United States have implemented similar pregnancy treatment programs through an m-health initiative. Medical practitioners schedule prenatal care visits for expectant mothers, provide remote advice during childbirth, and check in after childbirth to deal with any health issues.[21]

Wireless-enabled technology has aided quality-of-life measures and outcomes for many people. Forgetfulness is a common problem among patients, and forgetting to take medications is a particular and common manifestation, with serious results. It is estimated that only 50 percent of patients take their medication as prescribed.[22] Failure to take a drug at the time or in the dosage prescribed, as well as failing to take it entirely, means half the benefit of prescription drugs is lost through human error. This in turn costs the system billions in negative health outcomes. Research by the Telnor Group has found that m-health can reduce the costs of medical care among the elderly by 25 percent by reminding them to take their pills. These types of programs represent inexpensive ways to improve medical treatment and save money in the process.[23]

Extending Rural Access

Access to medical care in rural areas is a challenge in every country around the world. Nearly every nation has disparities between urban and rural areas. Health care providers and specialists are more likely to be located in densely populated jurisdictions because that is where hospitals and advanced equipment are found. While there are no guarantees that urban care is of higher quality than rural care, city hospitals in many countries have a richer array of imaging equipment, medication, and medical personnel.

Japan, an economically developed country, has a number of remote areas that are distant from urban centers. Some of them are located on sparsely populated islands or in mountainous areas far from major cities. To improve access to medical care, Medical Platform Asia gave the three hundred residents of rural Hokkaido 3G wireless devices that record and transmit blood pressure, weight, and distance walked through pedometers. Doctors examine the incoming data and make medical recommendations to each individual. Already there have been substantial improvements in patient awareness on issues such as blood pressure management (awareness went from 50 percent to 100 percent during the pilot project) and the importance of being proactive about medical care (engagement rose from 70 to 100 percent).[24] Awareness, of course, does not guarantee positive results, but it at least creates the possibility that people can identify certain health conditions and seek early invention when problems develop.

In India, a rapidly developing economy, rural dwellers gain access to medical care far from their hometowns through videoconferencing. Using broadband connections, doctors geographically remote from patients can examine them and diagnose particular problems. In countries where physicians are in short supply in rural areas, this enables those in underserved locales to get medical treatment.[25]

In Southeast Asia, Singapore has a mobile health application called Health Buddy that provides a list of medical symptoms and possible treatments, along with tips and videos on ways to promote good health. Patients can ask specialists detailed questions about particular illnesses and whom they should see for medical care. The app provides access to SingHealth, the government health portal with over forty different medical specialties available.[26]

In Malawi, Josh Nesbit of Medic Mobile has developed software that allowed health workers to text in medical information for rural patients. Rather than spend hours commuting to clinics, they can get quick diagnoses for routine symptoms and suggested treatments. According to him, "within six months of the system going live, the number of patients being treated for tuberculosis doubled, more than 1200 hours in travel time were eliminated, and emergency services became available in the area for the first time."[27]

Pilot projects in India and Sri Lanka have found wireless-enabled remote monitoring very helpful in identifying and tracking outbreaks of dengue fever. Before wireless communications were feasible, it generally took the provinces fifteen to thirty days to report data on disease outbreaks to central authorities. The long delay in reporting and receiving data slowed treatment responses and aggravated the spread of infectious diseases. However, with the availability of mobile and digital communications, the wait time to diagnose and treat infectious diseases has dropped considerably.[28] Through the Real-Time Biosurveillance Program, which incorporates mobile devices, public health authorities use data mining techniques to detect anomalies in disease patterns. Areas reporting major outbreaks receive additional resources for diagnosis and treatment, and this strategy helps limit the spread of the disease.

Mobile applications make doctors more efficient because they don't have to be in the physical presence of a patient to judge his or her condition: according to the Telnor Group, m-health can double access to physicians by those living in rural areas.[29] Digital technology allows people to overcome the limitations of geography in health care and access information at a distance. Patients are able not only to receive medical advice but also to get a second opinion without visiting another physician by sending that person relevant medical test results. If a personal conference is required, doctors can use videoconferencing to speak to patients located in another city or state.

Overcoming Policy Barriers

Many patients want to employ digital and mobile technologies in their health care. For example, 77 percent in a national survey conducted by the Pew Research Center said they would like to get reminders by e-mail from their doctors when they were due for a visit, 75 percent wanted the ability to digitally schedule a doctor's visit, 74 percent wanted to use e-mail to communicate directly with their doctors, 67 percent wanted to receive the results of diagnostic tests by e-mail, 64 percent wanted access to an electronic medical record to capture information, and 57 percent said they would like to use a home monitoring device that would allow them to e-mail blood pressure readings to their doctor's office.[30]

There are substantial concerns, however, about the privacy and security of data transmitted by mobile devices, and many of those interested in using wireless technology to manage their health wonder whether the increased use of handheld devices will compromise the confidentiality of their medical information.[31] They fear the loss of a smartphone or tablet that contains personal material and are concerned that their medical information might end up in the hands of employers or private individuals.

To move forward with mobile technology, providers must overcome both privacy concerns and policy barriers in terms of reimbursement, regulation, and research.[32] In many countries, neither public nor private insurance plans cover m-health applications. Physicians, for example, often are not reimbursed for e-mail or phone consultations, sending text messages to patients, or reviewing data gathered through remote monitoring devices. The reimbursement policy continues to favor face-to-face medical treatment over communications by digital or mobile devices. This historically based impediment limits physician interest in and use of innovative treatment approaches and reduces the benefits of the m-health revolution.

Thus, policy changes are needed that recognize the changing landscape of medical care and the benefits of remote monitoring devices, preventive medicine, text reminders to take medication, and electronic consultations. Unless physicians are reimbursed for these practices, they will be less likely to make use of new techniques.

The mobile revolution has also raised issues in terms of government regulation of medical devices and diagnostic tools. In the United States, the Food and Drug Administration (FDA) has responsibility for ensuring that medical devices are safe and beneficial. With prescription drugs and expensive imaging systems, the FDA requires clinical trials proving effectiveness and lack of harm through adverse conditions before it approves their use.

If patients rely on home monitoring devices to transmit blood pressure data, should there be an accuracy requirement in the data transmission? Or if there are mobile apps that enable physicians to read ECGs or CT scans on handheld devices, should governments regulate the app to ensure effectiveness?

In 2012, the FDA published draft guidelines stating that "the interpretation of imaging scans on a mobile device could be affected by the smaller screen size, lower contrast ratio and uncontrolled ambient light of the mobile platform."[33] Agency officials indicated that they would look at apps with an eye toward these considerations, and would also evaluate the marketing claims made on behalf of mobile applications and physician decision support protocols developed to help diagnosis and treatment.

One of the issues the FDA is grappling with is whether to regulate medical screening and diagnostic tools. In a *Federal Register* posting, the agency said it was "aware that industry is developing new technologies that consumers could use to self-screen for a particular disease or condition and determine whether a particular medication is appropriate for them. For example, kiosks or other technological aids in pharmacies or on the Internet could lead consumers through an algorithm for a particular drug product. Such an algorithm could consist of a series of questions that help consumers properly self-diagnose certain medical conditions or determine whether specific medication warnings contraindicate their use of a drug product."[34] The agency indicated that the use of these types of algorithms could save money because it would free consumers to self-diagnose common problems and therefore save unnecessary trips to a doctor's office.

Even as m-health grows and mobile devices and apps prolifer-ate, it is clear that more research is required to link the use of mobile technology to health outcomes. Considerable data are already available showing positive results on such metrics as user satisfaction, reductions in wait time, improved attendance at med-ical appointments, and significant cost savings. But more informa-tion is needed to establish a solid connection to health outcomes, such as drops in infant mortality, reductions in the spread of infectious diseases, and better management of chronic illnesses. Those are the ultimate objectives of health care, and researchers need to focus attention now more on outcomes.[35]

Moreover, m-health provision is limited by the need to over-come economic, organizational, and technology disparities across nations. Research by Patricia Mechael and colleagues at the Columbia University Center for Global Health and Economic Development has identified several sources of inequity: treatment compliance, disease surveillance, health information systems, point-of-care support, health promotion, disease prevention, and emergency medical response.[36]

These issues make it difficult for people in all nations to share equally in the benefits of the unfolding technology revolution. Countries that have made progress in developing m-health initia-tives should transmit their best practices to other nations so that the latter can understand ways to move forward. That would help more people unleash the potential of their mobile devices and gain the virtues of new wireless-enabled technologies.

The Economic Impact of m-Health

M-health clearly has expanded in number and type of initiatives. It is expected to become a multi-billion-dollar field by 2017. Accord-ing to a consulting report from PricewaterhouseCoopers, annual revenues are projected to reach $23 billion worldwide by 2017. This figure is apportioned to $6.9 billion in revenues expected in Europe, $6.8 billion in Asia, $6.5 billion in North America, $1.6 billion in Latin America, and $1.2 billion in Africa.[37]

Remote monitoring is expected to account for about two-thirds of this market as doctors and patients use mobile devices to manage chronic illnesses. As mobile devices become increasingly common in both developing and developed countries, applications to promote awareness, prevention, diagnosis, and treatment have multiplied. Mobile technology is especially helpful in managing chronic disease states because it frees physicians and patients from routine office visits while still providing data on patient conditions. This allows doctors to focus office care on those needing more detailed medical assistance.

A study of the U.S. wireless industry by Roger Entner found that mobile devices improve worker productivity in four ways: (1) by reducing unproductive travel time, (2) by improving logistics, (3) by enabling faster decisionmaking, and (4) by empowering small businesses and improving communications. He estimated that the industry increased productivity by $33 billion in 2011 alone. One-third of this gain ($11.2 billion) was attributed to the medical area. Entner projected productivity gains of $305.1 billion over the next ten years in medicine.[38]

Econometric modeling shows a significant tie between mobile penetration and economic development. Mobile communications contribute 0.39 percent to GDP growth in the developed world.[39] Others have suggested a bigger effect of 0.59 percent added to national growth.[40] By easing communication, improving service delivery, and reducing transmission errors, mobile devices contribute positively to economic growth and thereby benefit countries around the world.

7 MEDICAL DEVICES AND SENSORS

HEALTH CARE ACCESS, AFFORDABILITY, and quality are problems the world over. There are well-established disparities based on income and geography, and the high costs of health care present affordability challenges for millions of people. Large numbers of individuals do not receive the quality care they need.

Mobile technology offers ways to help with these challenges. Through mobile health applications, sensors, medical devices, and remote patient monitoring products, avenues for improving health care delivery exist. These technologies can help lower costs by facilitating the delivery of care and connecting people to their health care providers. Applications allow both patients and providers to have access to reference materials, laboratory test results, and medical records using mobile devices.

Complex mobile health applications are useful in such areas as training health care workers, managing chronic disease, and monitoring critical health indicators. They enable easy access to tools such as calorie counters, appointment notices, medical references, and physician or hospital locators. These applications empower patients and health providers to address medical conditions proactively, through near real-time monitoring and treatment, no matter the location of the patient or provider.

In this chapter, I look at specific applications and inventions and discuss how mobile technology is transforming health care in the United States and around the world. I propose that mobile health care devices and applications help front-line health care workers and care providers be more efficient and effective in their provision of medical assistance. Finally, I recommend several steps to speed the adoption of mobile technology in health care.[1]

Innovations in Mobile Health Care

The health care area is undergoing a technology revolution. New services are being offered online, and consumers can use mobile phones to connect to a variety of different providers. For example, applications such as the iWander app for Android devices are being used for patients with Alzheimer's disease or dementia. People use the GPS function of smartphones to track patient locations. If the individual travels away from home or other known locations, it triggers a signal to the person's family or caretaker to check on the individual's status. Through geolocation coordinates, the person can easily be found and returned to the care setting.[2]

Social media sites are also helping patients cope with specific diseases. For example, diabetes-related complications represent a major source of emergency room visits. A study on the TuDiabetes.org site asked patients to report their experience with hypoglycemic events, age, gender, use of insulin pumps, and health issues. On average, the approximately 500 respondents reported experiencing six insulin-related problems.[3] By sharing experiences, viewers could see what others had experienced and learn ways to cope with particular health emergencies.

Some of these applications have been developed for the cloud, which bypasses issues of compatibility of devices and apps across platforms. Interoperability challenges in the health care field manifest in numerous ways. In some cases information systems are not able to communicate with one another. At other times there are incompatibilities in terms of data files, semantics, or file-sharing protocols. Placing the wireless solution on a cloud storage system moots connectivity issues and makes it easier to communicate across different information regimes.

One system is the 2net Platform, developed by Qualcomm Life. This cloud-based system, designed to be compatible with different kinds of medical devices and applications, transfers, stores, converts, and displays data acquired by electronic medical devices.[4] Both patients and care providers have access to the information around the clock. Aggregating data in a single, accessible location is a tremendous benefit, especially during medical emergencies.

With the development of many new mobile applications, it is important to determine which ones are most helpful. iMedicalApps.com is the leading outlet for medical personnel. The analysts who post on the site offer reviews and commentary on mobile medical technology. Readers can see which applications receive the most positive reviews and which ones do not. The site also offers recommendations on the top health care applications available on the market.

Remote Testing and Diagnosis

There has been growth recently in "wearable sensors" and remote monitoring devices. Some of these products have been listed with the FDA, while others are off-shore devices that are not for sale in the United States. Among the devices available in the United States is a portable electrocardiogram (ECG) system for high-risk cardiac patients. The system uses smartphones attached to heart monitors to transmit heart rhythm data to health care providers. Software analyzes the ECG waveforms for possible abnormalities.[5] Patients needing special attention because of anomalies are notified of possible problems that should be addressed.

Propeller Health has developed an inhaler with an asthma sensor built into it. The sensor tracks environmental conditions that pose possible dangers to asthma sufferers. By keeping track of external conditions as well as how often the person is taking medicine, the device helps patients manage asthma and keeps health care providers informed about disease management.[6]

AirStrip was cofounded in 2005 by the web developer Trey Moore and the Texas obstetrician Dr. Cameron Powell. It uses the wireless software AirStrip OB to monitor fetal heart rhythms and send data on pregnant women to their physicians. The entrepreneurs have also developed AirStrip Cardiology, which monitors cardiac data, and AirStrip ICU, which monitors data acquired from intensive care or emergency room patients. AirStrip's research arm has found improvements in physician satisfaction following the deployment of these mobile health applications. Surveys of obstetricians found that "mean overall satisfaction scores are 4.5 to percent 10.9 percent higher at hospitals using the AirStrip applications than [in those hospitals'] regional and national peer group counterparts."[7]

Research by Suneet Chauhan and colleagues has demonstrated the value of electronic fetal monitoring. In a study that looked at nearly 2 million infant birth and death records, he and his colleagues found a 53 percent reduction in mortality when electronic fetal heart rate monitoring devices were used.[8] Having real-time data on possible abnormalities or developing conditions helps health care providers identify who is at risk and what can be done to prevent the onset of more serious conditions.

The Turkish researchers Oguz Karan, Canan Bayraktar, Haluk Gumuskaya, and Bekir Karlik have combined smartphone technology with software algorithms to create "pervasive healthcare services."[9] Their system allows users to enter data into their smartphones dealing with age, physical activity, pregnancy status, body mass index, skin fold thickness, cholesterol level, diastolic blood pressure, serum insulin level, plasma glucose concentration, as well as whether diabetes runs in the family. Along with personal medical history, this information is transmitted in real time to health care providers with decisionmaking support that tells them whether the readings are normal or abnormal. Those with abnormal signs are advised to seek medical assistance.

Zephyr is a firm founded in 2003 that offers a heart rate monitor enabled through a mobile device for use by consumers, soldiers, first responders, and athletes. It tracks heart activity, respiratory rate, ECG signals, stress levels, posture, activity level, and peak acceleration. The tool straps onto the chest and records data on physiological status. It gives consumers and professionals a powerful tool for monitoring their own health status.[10]

These products represent just a few of the new services and monitoring devices designed to help people manage particular illnesses. A great number of innovative approaches have come on the market. With health care systems needing to improve access, affordability, and service delivery, inventors have turned to mobile devices to provide better health care.

Empowering Front-Line Health Care Workers with Medical Knowledge

In many places around the world, front-line health workers have difficulty accessing medical information or learning from the experiences of health colleagues. Often, they don't have common medical reference materials or basic knowledge about diagnosis, treatment, and prescriptions.

A project called mPowering Frontline Health Workers is addressing this problem by using mobile devices to provide the latest medical information to front-line health care providers. Through a digital repository provided by health experts, care providers such as midwives, nurses, and community health workers can use cell phones, smartphones, tablets, and laptop computers to obtain information on neonatal care, immunization, and childhood diseases. This ability to access knowledge helps them be more effective in delivering health care and reducing maternal and child mortality in developing nations.[11]

In South Africa, for example, health care providers use mobile devices with a library of clinical resources. Nurses and physicians can access the latest medical information concerning diagnosis, treatment, and medication. They can look up data on drug interactions or ways to treat a particular illness.[12]

In Japan, the Wireless Health Care@Home program allows residents living in rural areas to send critical health information to doctors via a wireless network. Giving people the means to better manage their own health while also receiving timely treatment helps prevent illnesses from becoming more serious.[13] It is a way to empower patients by giving them greater responsibility for their own medical treatment.

These are just some of the ways that mobile devices improve health care by putting timely and up-to-date information at people's fingertips. If health care providers are able to check the adverse side effects associated with certain treatments or research how particular medications affect patients with a given disease, tremendous benefits will be realized by the health care system. Such mobile-based capabilities help reduce the costs of care while also improving the quality of medical care delivered.

Facilitating Mobile Health Innovation

Invention has aided the development and deployment of the applications and systems described here. Those who build medical devices and develop software applications need an environment that encourages discovery and creation. This includes a culture that facilitates invention and rules that help inventors make money from their various creations.

Medical device inventors Dr. Howard Levin and Mark Gelfand have described how they operate.[14] Levin is a heart transplant cardiologist, while Gelfand is a systems engineer. In thinking about health problems, they start with a list of up to thirty ideas and then whittle it down. Using fellow experts and the medical literature, they analyze factors such as whether the product would meet an unmet need, pose a technical risk, be used on patients, and produce any adverse side effects.

In the United States, there are in addition a number of issues that have to be addressed regarding government regulation of mobile medical devices. One is the question of whether to regulate particular products. Some devices are marketed for health and fitness monitoring and therefore are not subject to device regulation. Calorie counters and activity monitors fall in this category. As consumer items that have no discernible risks and are noninvasive, there is no reason for the U.S. Food and Drug Administration to oversee them.

Devices that are thought to pose some patient risks are subject to regulation. The FDA's Center for Devices and Radiological Health groups 1,700 medical devices into 16 medical specialties. It focuses specifically on "radiation-emitting electronic products" used in diagnosing or treating diseases in the United States. Companies must register as manufacturers and list their products, and potentially need clearance through a premarket notification 510(k) process or premarket approval, in addition to meeting appropriate labeling requirements, instituting good manufacturing practices, and reporting adverse events through the Medical Device Reporting system.

Each device is assigned to one of three regulatory classes based on intended use and possible risks to patients.[15] Stethoscopes are an example of a Class I device. Devices in Class I are subject only to general controls since they pose lower risks to patients. Class II devices such as scanners are considered higher risk, and general controls alone are insufficient to provide reasonable assurance of safety and effectiveness; such instruments require premarket notification through the 510(k) review process. Brain stimulators and cardiac defibrillators are examples of the highest-risk devices, ones that support or sustain human life and are of substantial importance in preventing impairment of human health or represent a potential, unreasonable risk of illness or injury. Class III devices call for clinical studies demonstrating safety and effectiveness.[16]

Other countries vary considerably in how they handle medical devices. The European Union, for example, allows the marketing of mobile devices across all member countries once they have been approved as Conformité Européenne in any one of the EU member nations. It uses four categories of devices, Class I, IIa, IIb, and III, based on risks and intended use.

Medical devices are approved if "the device successfully performs as intended in a manner in which benefits outweigh expected risks."[17] Class III devices in Europe require clinical trials, but the details of the trials are not made public and are not binding on manufacturers. Information on serious adverse events must be reported to the relevant government authority but are not publicized to the general public.

Some commentators have expressed concern that European regulators are paid directly by device sponsors and that they focus most on whether medical tools work as intended, as opposed to their impact on public health. The medical researchers Daniel Kramer, Shuai Xu, and Aaron Kesselheim say the European process is faster and requires less detailed clinical studies. They claim European regulators have approved monitoring devices for coronary artery interventions based on twenty-two subjects, compared to eight hundred people for the same device in the United States.[18]

Others have suggested the value of postmarket oversight of medical devices. Under current law, patients, physicians, and manufacturers convey device failures or adverse impacts to the U.S. Medical Device Reporting system. Although that database receives more than 100,000 complaints each year, fewer than 0.5 percent involve medical device failures.[19] Yet with the FDA approving more and more devices through substantial equivalence of predicate device in the 510(k) clearance pathway as opposed to premarket approval requiring detailed analysis, these authors feel the need for better oversight.

Still others have expressed concern about the level of medical device user fees. The FDA first started collecting fees in 2002. Device manufacturers pay fees when they register as manufacturers and list their particular devices for marketing with the FDA. Assuming that Congress does not kill the device fee as part of a budget agreement, the federal agency plans to collect around $595 million between 2012 and 2017 through this means.[20]

The FDA recently published its final "guidance" on mobile medical apps that outlines its "current thinking" and approach to mobile medical applications.[21] The guidance defines mobile applications and mobile platforms and explains what is subject to agency oversight. For example, government officials do not include electronic copies of medical textbooks or reference materials as subject to regulation. Apps that distribute educational materials for patients or provide suggestions regarding general health and wellness are not included. The same is true for mobile apps that help with general office operations such as billing, appointments, medical claims, business accounting, medical reminders, or insurance reimbursements.[22]

The types of mobile devices it seeks to regulate include those that

—"use a sensor or lead that is connected to a mobile platform to measure and display the electrical signal produced by the heart,"

—"use a sensor or electrode attached to the mobile platform or tools within the mobile platform itself (e.g., microphone and speaker) to electronically amplify and 'project sounds associated with the heart, arteries and veins and other internal organs,'"

—use sensors to measure "physiological parameters during cardiopulmonary resuscitation,"

—"analyze eye movements for use in the diagnosis of balance disorders,"

—use sensors that examine "degree of tremor caused by certain diseases," "electrical activity of the brain," "blood oxygen saturation," or "blood glucose levels,"

—connect "to an existing device type for purposes of controlling its operation, function, or energy source," or

—"transform a mobile platform into a regulated display, transfer, store, or convert patient-specific medical device data from a connected device."[23]

Finally, for a variety of mobile apps the FDA said it would exercise "enforcement discretion (meaning it will not enforce requirements under the Federal Food, Drug, and Cosmetic Act) for the majority of mobile apps as they pose minimal risk to consumers."[24] This category included such items as the following:

—devices that "provide periodic educational information, reminders, or motivational guidance to smokers trying to quit,"

—devices that "use GPS location information to alert asthmatics of environmental conditions that may cause asthma symptoms," or

—"mobile apps that use video and video games to motivate patients to do their physical therapy," and

—mobile apps "that aggregate and display trends in personal health incidents."[25]

Overcoming Barriers

Mobile technology offers interesting ways to help with health care access, affordability, and service delivery. Through mobile applications, sensors, remote monitoring devices, and reference materials, there are numerous avenues by which health care delivery can be improved. Invention is a big part of this ecosystem because it is difficult to build new hardware or construct software applications without a broader environment that encourages and rewards inventiveness.

To encourage mobile health, there are several actions designed to improve the adoption of mobile medical devices and applications. Policymakers should encourage the use of mobile devices for health care. Smartphones and tablets have spread rapidly in developed and developing nations, and this represents a major opportunity to transform the manner in which medical care is delivered. Moving to electronic systems for service delivery will save money, improve access, and provide higher quality of care.

Chronic diseases are a costly part of the current system. Nearly three-quarters of medical expenditures take place for a small number of chronic illnesses, including cardiovascular disease, cancer, diabetes, and asthma. We should encourage the use of mobile systems that monitor patient symptoms and provide real-time advice on treatment and medication because they have the potential to control costs, reduce errors, and improve patients' experiences.

There now are mobile applications that aid in chronic disease management, sensors and remote devices that monitor patient physiology, and electronic libraries that bring the latest knowledge to health providers around the globe. These materials represent a quantum leap forward in offering quality health care. We should work to remove barriers to adoption and make these tools much more widely available.

The same is true for clinical decision support for health care providers. With growing knowledge about diseases, genetics, and pharmaceutical products, the practice of medicine has become far more complicated. Physicians are expected to know the latest advances in medicine and apply that information to their patients. Software that helps health providers understand how to deal with particular symptoms and what drug interactions should be avoided is increasingly viewed more as a reference library than as a medical device, calling into question how it should be regulated. Health providers need access to as much accurate data as they can get on how to treat various ailments. In addition, one of the barriers to cost containment and quality service delivery has been the continued reliance in many locales on paper-based medical systems. Physicians prescribe medicine through manual forms, laboratory tests are reported on paper or film, medical records are stored in filing cabinets, and insurance claims are paid through reimbursement requests sent through the mail. In a digital world, one cannot imagine a costlier way to run a health care system.

On the issue of government regulation, the FDA has finalized its guidance on how mobile applications and regulated mobile medical devices are to be treated, in an effort to clarify some of the ambiguities and help further innovation. Having clear rules that encourage desirable behavior is the best way to move forward in mobile health.

Many countries, including the United States, are challenged to provide adequate health care. Difficulties include physical distance between doctors and patients, too few skilled health care professionals, and problems controlling health care costs and funding new medical infrastructure. In addition, the current epidemic of chronic illnesses, in both developed and developing economies, illustrates the need for innovative, efficient, technology-supported interventions.

Mobile technologies offer the ability to connect patients with their doctors, caregivers, and loved ones and to enable timely health monitoring, all of which have the potential to improve patient engagement and health outcomes. Such technologies can aid in providing access to information, helping to lower costs, facilitating remote care, and increasing efficiencies by connecting patients to their providers virtually anywhere. As their implementation spreads, these applications and services are becoming an essential tool in extending health care resources around the world.

8   PUBLIC OUTREACH

DURING THE CAMPAIGNS LEADING up to the 2008 and 2012 presidential elections, candidate Barack Obama pioneered innovative uses of digital technology. With the help of the Internet, he raised hundreds of millions of dollars. He used social media platforms such as Facebook and Twitter to identify and communicate with supporters around the country. And through videoconferencing, he launched virtual get-togethers with voters in many different locales simultaneously.

In so doing, he helped to promote the campaign use of mobile technology and transformed campaigning and public outreach. Candidates, voters, activists, and reporters now are using hand-held devices for fundraising, field organization, political persuasion, media coverage, and government accountability. Recent elections have seen the widespread utilization of mobile ads, video, web links, and apps.

In this chapter, I review innovative examples of campaign outreach made possible through mobile technology. I show how smartphones expand the opportunities for mobilization and provide new ways to connect people and leaders. With some policy changes, it would be possible to use the affordances of mobile technology to expand citizen engagement in the political process.[1]

Public Outreach and Voter Engagement

The use of smartphones for outreach and voter engagement has risen dramatically in recent years. According to Jumptap Founder Jorey Ramer, "consumers are spending 10% of their time with mobile media."[2] They like the convenience, immediacy, and ability to personalize communications through handheld devices. Scott Goodstein, the CEO of Revolution Messaging, explained that "whether it is through in-app, MMS, a mobile landing page or a YouTube clip, the ability to watch a video through mobile makes a direct connection between candidates and voters."[3]

With surveys indicating that 83 percent of smartphone and tablet users are registered to vote, it is no surprise that American candidates are using these devices to identify voters and recruit volunteers.[4] According to a survey by the Pew Internet & American Life Project, "26 percent of Americans used their cell phones to connect to the elections." Twelve percent indicated they used cell phones to keep up with the news, 10 percent sent election texts to friends and family members, 6 percent used their phones for information about voting places and conditions, 1 percent relied on mobile devices for election-related apps, and 1 percent contributed campaign money through text messaging.[5]

Candidates reached out through mobile apps, mobile advertising, and text messaging. For example, in the 2012 Republican presidential primaries, Newt Gingrich asked interested voters to text "Newt" to 59769 "to volunteer and get text updates on helping Newt win."

Other politicians are developing mobile apps that take viewers to campaign websites. For example, President Obama's campaign developed an app that allowed people "to access photos and videos, receive news updates and donate to the campaign."[6] Republican nominee Mitt Romney also employed mobile apps to engage voters and attract supporters.

These outreach strategies are not limited to the United States. Mobile devices are being used in places the world over to engage voters. Candidates and political parties are bypassing traditional media and going directly to voters for political persuasion. Although some of the emphasis on communicating by mobile technology is the result of lack of landlines or civil unrest, developed nations enjoying civil peace are well represented among those using direct-to-voter apps. In Germany, for example, the Berlin Green Party developed a mobile app that "allows supporters to discuss environmental issues around the city and brings party billboards to life using Augmented Reality."[7] When pointed at a billboard, the app launches a mobile video message that discusses the environmental issue in greater depth.

In the United Kingdom, university students developed an app called PoliticsDirect that "locates the mobile user's geographical location to find their local MP, MEPs, Councillors, Council and relevant information about them such as voting record and expenses."[8] This helps voters hold public officials accountable and improves the responsiveness of the electoral system.

Field Organization

Field organization is key in any election, and candidates are using smartphones to identify likely voters and turn them out on election day. For example, in his Massachusetts race for a U.S. Senate position, Scott Brown used smartphones with GPS to help field canvassers identify undecided voters. Workers were able to walk down each street, determine who was undecided based on canvassing calls, and knock on the doors or place phone calls to relevant individuals.[9]

The Obama campaign allowed volunteers to log on to the Obama website with their Facebook ID and get access to "any tool that you can get in a field office. You can have that at home, on your computer, in real time, in a way that connects to what your friends are doing and what the people around you are doing," according to chief integration and innovation officer Michael Slaby.[10]

This enabled voter outreach because "people can now make calls, canvass, and be engaged on a deeper level from wherever they are," Slaby said. By integrating mobile devices, tablets, and personal computers, campaign organizations "[made] it easier for volunteers to register new voters and call undecideds on the go."[11]

Candidate Romney followed a similar approach. According to his digital director, Zac Moffatt, "We're going to have people watch one of our rich video units online, engage with our campaign, syndicate the message through social sharing, vote for us, and convince their friends to do the same."[12]

In Haiti, where 75 percent have mobile phones, voters use these devices to "access information, confirm voter registration status and obtain information on the location of polling centers." This has helped the country organize elections in the aftermath of its devastating earthquake.[13]

Public opinion research in the Czech Republic found that 55 percent of the country's nonvoters said they would cast a ballot if they could vote via mobile phones. This means that an extra 500,000 individuals said they would vote if they could do so with the convenience and ease of using their personal cell phone.[14]

Advertising and Political Persuasion

With the growth of mobile usage, political organizations have turned to online advertising as a way to persuade voters. For example, people who attended the Minnesota State Fair and had a smartphone received targeted ads on their mobile devices from Michele Bachmann's congressional campaign informing them that her opponent supported food tax increases. "I know it's State Fair time and you don't want to hear about politics," the ad announced. "But while you're at the Fair, you should know that Tarryl Clark here voted to raise taxes on your corn dog and your deep-fried bacon and your beer."[15] The campaign was able to target this ad only to those individuals who were within a two-mile radius of the state fair.

Geolocation features enable candidates to target ads geographically to air during specific events. For example, voters attending the Iowa caucuses in 2011 received mobile ads targeted to caucus-goers. Candidates have the same capability to reach individuals who attend specific speeches or campaign rallies. "Campaigns want to reach voters where they are," indicated Rob Saliterman of Google Advertising. "And because of that, I think we'll see more [mobile advertising]."[16]

Voters have received "geotargeted" mobile ads when they attended a Texas Rangers World Series game (sponsored by an Arlington, Texas, congressional candidate endorsed by Rangers' owner Nolan Ryan), a NASA shuttle launch (sent by a Florida Senate candidate claiming that incumbent Senator Bill Nelson supported ending the space program), or the University of Virginia (originated by former U.S. representative Tom Perriello, hoping to mobilize the youth vote).

When he campaigned in South Carolina, Republican Rick Perry used mobile advertising to target nine Christian colleges in an effort to mobilize evangelical voters.[17] Meanwhile, the Mitt Romney campaign employed Google's mobile search Adwords to engage voters. Using a "click-to-call" feature, viewers who searched for "Mitt Romney" would see an ad that allowed them to click through and call a Romney campaign office for more information.[18]

One of the reasons that candidates like mobile advertising is its highly targeted nature. Candidates can target Zip codes, area codes, or other geographic areas. "Mobile apps create a 1 to 1 message directly to your user. It's like a personal conversation. You have the ability to control the worth of your mobile app user conversation. Mobile apps are quite engaging, and engagement is a key to business," said Cami Zimmer, the CEO of Campaign Touch, a mobile communications firm.[19]

According to Insight Express research, mobile campaigns are very effective at raising awareness. Its study found that "mobile Internet campaigns resulted in increases of 9 percentage points for unaided awareness, 9 percentage points for aided awareness and 24 percentage points for ad awareness."[20]

Fundraising

A U.S. survey conducted by the Pew Research Center's Internet & American Life Project discovered that 9 percent of respondents (Americans) said they had used mobile phones to text $10 charitable contributions using the word "HAITI" to the phone number 90999.[21] After the Haiti earthquake in 2010, $43 million was raised by text donations. For 74 percent of the donors, the Haiti campaign was the first time they had made a charitable contribution using a mobile device, and three-quarters of them did so as a spur-of-the-moment gesture, without undertaking much research.

After seeing how easy it was to contribute by smartphone, a number of donors made gifts to other disaster efforts. For example, people can give money to UNICEF by texting "FOOD" or to the International Rescue Committee by texting "AFRICA."[22] Overall, 40 percent said they had texted a gift after the 2011 Japanese tsunami and earthquake, 27 percent did so after the 2010 British Petroleum oil leak in the Gulf of Mexico, and 18 percent gave a mobile gift to help those harmed by U.S. tornadoes during 2011.[23]

Until recently, however, federal candidates in the United States were forbidden by law from accepting political contributions by text messaging. In a 2010 advisory ruling, the Federal Election Commission (FEC) rejected a CTIA—The Wireless Association request to allow $10 text contributions for federal candidates. The FEC's rationale was that mobile texting would undermine finance disclosure rules if individuals could text contributions without having to provide their name, address, and occupation, as required by federal law, and that contributing by text might allow foreigners or corporations to provide money illegally.[24]

However, since disclosure is required only for contributions over $200 and since donors can anonymously provide donations up to $50 a person, this policy decision has been reversed. The FEC now allows text donations of up to $50. The goal is to encourage small donors to participate in the finance system and provide a counterweight to the large amounts of money entering the political process from wealthy individuals and corporations. Donors must affirm that they are not foreigners or corporations, neither of which are allowed to make direct contributions.[25]

Scott Goodstein has argued that wireless carriers need to reform their billing practices to make sure donors do not exceed the $50 limit for anonymous political contributions and that they forward contributions within ten days, as required by federal law. Reliance on monthly billing cycles is not sufficient for political campaigns, he says.[26]

The Obama campaign accepted campaign contributions through Square mobile credit card readers. This mechanism allowed campaign workers to accept political donations sent to their iPhone or Android devices. Reliance on the use of card readers expedites the processing of contributions and enables field staff to raise money from contributors around the country.[27] The campaign developed a custom app that, when used in conjunction with Square, collected FEC-required information, such as address, occupation, and employer.[28]

The states of California and Maryland have approved legislation that allows supporters to text contributions to political candidates.[29] California law does not limit contribution amounts, but Maryland legislation places a $10 limit on political contributions, in accordance with the preferences of most wireless carriers. Carriers prefer the contribution amount to be limited because the gift is added to the user's monthly phone bill, which can be contested.[30]

Some countries allow political party donations to be made through mobile devices. In South Africa, for example, supporters can text contributions to the nine parties that have signed up for the service. However, the payment service provider charges 5 percent of the contribution to process the gift.[31] In the United Kingdom, there are mobile apps that allow people to make a political contribution by texting to the party name, such as "GREEN."[32]

Media Coverage

Mobile technology is affecting how reporters cover politics. A variety of news organizations are deploying election apps designed to improve coverage and provide voters with the latest campaign news. For example, the *New York Times* launched a 2012 election app that provided "news, polling data, candidate information and—when the time comes—live election results."[33] Not only did the app include *Times'* stories, it relied on coverage from other sites that the editors deemed important, such as *The Economist* and CQ Roll Call.

Government Accountability

Mobile technology is a powerful tool for improving political accountability. Work by Charles Gibson and his colleagues at the University of California at San Diego demonstrates that cell phones have helped reduce electoral fraud in Uganda and Afghanistan. For example, people armed with cell phones monitored electoral processes in Afghanistan and reduced electoral fraud by 25 percent for politically connected individuals and 60 percent for the theft of ballot materials. In Uganda, where analysts studied 1,002 polling places, fraud was decreased by 26 percent when people used cell phones to take pictures of posted local election tallies and send them to a central clearinghouse.[34]

Improvements in electoral accountability and fraud reduction also have been found through mobile monitoring in other countries. A study of the nonprofit National Election Watch in Sierra Leone found that the combination of citizen monitors and cell phone communications helped safeguard the electoral process. Observers sent codes via text messages to National Election

Watch headquarters, which then aggregated the results. Within eighteen hours of poll closing, the organization announced that "despite minor discrepancies in certain districts, the elections had been well run by the election commission and voters had turned out in large numbers."[35]

Similar results were found in elections that took place in Indonesia, Palestine, and Montenegro. Election and poll monitoring through cell phones improved the conduct of vote tabulation. Observers could see what was happening in various districts around each country, report quickly to a central headquarters, and identify places that had problems. The immediacy and real-time nature of the observation deterred bad behavior and helped to ensure quality elections.[36]

The examples proferred in this chapter underscore the innovative uses of mobile technology in election campaigns around the globe, especially in relation to public outreach, fundraising, field organization, political persuasion, media coverage, and government accountability. Candidates, parties, voters, and reporters are using smartphones and mobile devices for text messaging, to send mobile ads, to provide videos and web links, and to draw users to election-related apps. These features have enriched the civic discourse and provided new opportunities for public outreach and citizen engagement.

9 DISASTER RELIEF AND PUBLIC SAFETY

THE GROWING USE OF mobile technology has greatly improved disaster relief and public safety efforts. Countries around the world face threats from natural disasters, climate change, civil unrest, terrorist attacks, and criminal activity. Mobile devices, tablets, and smartphones help emergency providers and the general public manage these challenges and mitigate public safety concerns.

This chapter focuses on how mobile technology provides an early warning system, aids in emergency coordination, and improves public communications. In particular, I review how mobile devices assist with public safety, disaster planning, and crisis response. I explain how these devices are instrumental in the design and functioning of integrated, multilayered communications networks. I show how they have helped save lives and ameliorated human suffering throughout the world.

Early Warning Systems in Natural Disasters

With the onset of global warming and climate change, tsunamis, flooding, drought, hurricanes, and tornadoes are becoming more

This chapter was co-written with Elizabeth Valentini.

common, if their onset is still unpredictable. Further, according to an overview of weather changes by the Intergovernmental Panel on Climate Change, the number of unusually cold days and nights has decreased since 1950, while the number of unusually warm days and nights has increased.[1] These temperature changes are linked to an increase in heavy precipitation events and a trend toward extreme coastal high waters. Furthermore, data suggest that some regions of the world have experienced more intense and lengthier droughts.[2] Although the data are still being analyzed, observations gathered over the past fifty years suggest a dramatic rise in extreme weather conditions and events.[3]

These shifts are inflicting tremendous costs in terms of human life, injuries, and property damage.[4] As the population density along many coasts increases, storms have destroyed homes, upset global supply lines, injured large numbers of people, and cost countless human lives.

In response to natural disasters such as Hurricane Katrina (2005), the earthquake in Haiti (2010), the earthquake and tsunamis in Japan (2011), and the Oklahoma tornadoes (2013), mobile invention and application have skyrocketed. Mobile development has surged in reaction to the increase in need for instant and accurate information. In Australia, for example, researchers have created software that combines voice over Internet protocol technology with wifi to enable communication between mobile devices in areas where there is no available reception.[5] Such a feature is of critical importance in instances in which traditional communications networks are knocked offline for extended periods of time during or after a major crisis.

Similarly, in response to the 2011 earthquake and tsunami in Thailand, Japanese developers have created Aerial 3D. Using laser beams, the system projects small, luminous dots to create text in local spaces.[6] This provides emergency response information to people in need of help and allows them to use mobile devices to pinpoint their locations.

Corporations are also supporting the invention effort. In 2013, AT&T launched a public safety hackathon challenge that invited mobile inventors to submit applications meant to ease the work of first responders after a disaster. Through this competition came InstantAct, an application that provides public safety officials with an exact field location during disaster and a more robust, dependable way of communicating with victims by voice.[7]

In the aftermath of the devastating 2010 earthquake in Haiti, Google created a crisis response team to deliver tools to help those affected by disaster. The company's designers put together tools for use during the earthquake and tsunami in Japan. Its main search page displayed a tsunami warning and its search engine's Person Finder app allowed people to tell friends and family they were safe.[8]

With a focus on preparation, Sesame Workshop and Qualcomm, through its Wireless Reach Initiative, have launched a mobile safety program in China that uses the capabilities of mobile technology. The initiative uses a mobile device to help young children and their families learn how to deal with various kinds of emergency situations. It promotes an Android application and a mobile website with a chalkboard that gives young children experience "writing their names and addresses, dial[ing] their home phone numbers and draw[ing] pictures of their homes or emergency meeting places."[9]

This program incorporates Sesame Street content, including videos that help young people learn about emergencies and ways they can protect themselves. A pilot project has received positive responses from people living in the southern province of Guizhou. Participants say it has raised awareness, promoted crisis skills, and helped parents teach their children disaster planning and amelioration. This service is now available nationally for free over China Telecom's mobile network.

In the weeks following the Japanese tsunami, downloads of mobile disaster applications soared around the world. Many of

these applications were built using various Japanese early warning applications, including Yurekura Call and Japan AED Map,
as models.[10] Apple featured a new section in its App Store called
"Stay in Touch," which provides a number of disaster relief applications, such as the American Heart Association's Pocket First
Aid & CPR; QuakeWatch, which tracks earthquakes and sends
warnings using U.S. Geological Survey (USGS) data; Disaster
Alert, which provides information on instant global "active haz-
ards"; the American Red Cross's Shelter View app, which helps
users locate a nearby shelter; and the Emergency Radio app.[11]

A number of government agencies, such as the U.S. Department of Health and Human Services (DHHS) and global netizens, have compiled lists of helpful disaster relief tools. David
Burns of *Campus Safety Magazine* broke down a list of disaster applications into useful categories: (1) reference materials,
such as the applications First Aid, WISER (Wireless Information
System for Emergency Responders), and FEMA's mobile app
and Emergency Survival Handbook; (2) personal preparedness,
such as Survive Now, ICE (In Case of Emergency), BuddyGuard;
iTsunami, and IMPrepared, and (3) situational awareness,
including the USGS's mobile app Floodwatch and the American
Red Cross's Shelter View.[12]

The DHHS website features disaster applications as well as
mobile-optimized web pages (or pages easily viewed on a user's
phone), including disaster medicine tools such as the American
Red Cross's First Aid, the American Heart Association's Hands-
Only CPR, and REMM (radiation emergency medical management), which provides information on the treatment of radiation
and nuclear injuries and emergencies.

The page includes disaster-related resources such as SOS, by
ARC, which offers step-by-step videos for common emergency
situations; the American Red Cross's Hurricane, Tornado, Earthquake, and Wildfires applications, which provide real-time information on these natural disasters; and MyMedList, which allows

a user to program his or her medications into a virtual list for first responders. There is also information on hazardous events and weather, such as the National Weather Service's mobile-enhanced web page, and on disease outbreaks, such as the Outbreaks Near Me application and the CDC's FluView, an application that tracks influenza illness across the United States.[13]

Communities around the world have adopted mobile technologies for disaster relief coordination. The Bangladeshi government announced that tens of thousands of mobile users in flood- and cyclone-prone areas would receive advance warning of impending natural disasters via mobile alerts. These alerts, unlike SMS texts, which are delivered to users' inboxes, flash directly on the screens of mobile devices, ensuring that more users see the message.[14]

In Britain, the London Fire Brigade (LFB) launched the world's first 999 emergency Twitter feed in December 2012. According to the LFB press release, text-based communications surpassed traditional phone calls and in-person meetings in 2012 as the most frequent form of communication between adults in the United Kingdom. That development, combined with the fact that over 2,000 tweets are sent every second worldwide, made the 999 emergency feed a very appealing and effective emergency option.[15]

Moreover, mobile devices have become increasingly important in the developing world, facilitating communication between local residents, government officials, and first responders. Many applications provide important information in the areas of health, agriculture, disaster relief, and crime. Mobile broadband is growing dramatically in emerging nations. According to the World Economic Forum, it will grow from 61 percent of broadband connections in 2011 to 84 percent in 2016.[16]

It is notable that there were fewer than 20 million fixed-line phones across Africa in 2000, but by 2012 there were nearly 650 million mobile phone subscriptions.[17] Mobile devices are crucial in providing assistance to those in need by allowing extended families to assist from afar. Members of Zimbabwe's British diaspora, for instance, can visit the website Mukuru.com to order goods such as petrol for their loved ones back home. Recipients in Africa are texted a code to their mobile devices, which they show the petrol station to receive their goods.[18]

In East Africa, livestock herders use mobile phones to send early drought warnings, in an attempt to skirt disastrous agricultural calamities such as the drought that struck the Horn of Africa in 2011.[19] By November of that year, the European Union had allocated more than €705 million to the Horn of Africa.[20] In a 2013 statement, UN Food and Agricultural Organization official Robert Allport reported that "cellular phones eliminate delays in receiving field data, since all the information is relayed via mobile network. In addition, the information is assigned a geographic location, so locations are extremely accurate and available in real-time."[21]

Mobile devices have become paramount in Africa's efforts to better prepare for—and respond to—natural disasters such as extreme drought. And—of importance—mobile technology has put action directly into the hands of African citizens. A few years ago, two UN officials, one in London and the other in Nairobi, received the following text message: "My name is Mohammed Sokor, writing to you from Dagahaley refugee camp in Dadaab. Dear Sir, there is an alarming issues here. People are given too few kilograms of food. You must help." Using the web at a local Internet café, Mr. Sokor found the officials' numbers and solicited their help directly.[22]

Terrorist Attacks and Domestic Shootings

It is not just natural disasters that pose problems for community planners. Terrorist attacks such as the events of 9/11 and domestic crises such as the 2012 Sandy Hook Elementary School shooting in Newtown, Connecticut, have brought public safety to the forefront of first responder and law enforcement concern, particularly in educational settings.

Two days after the Virginia Tech shootings in 2007, North Carolina attorney general Roy Cooper established the Campus Safety Task Force, charged with analyzing the events of school shootings like that at Virginia Tech and preparing a report on lessons learned. The January 2008 *Report of the Campus Safety Task Force Presented to Attorney General Roy Cooper* drew on three task force meetings, which reviewed testimony from more than thirty experts, law enforcement officials, emergency management experts, campus administrators, educators, emergency responders, victims' advocates, and psychologists. In addition, the team analyzed extensive literature reviews and the results of a web survey of 110 public universities, community colleges, and private instructional institutions.

The committee found that fewer than one in ten campuses were equipped with a siren system or a campuswide public address system. The report urged the adoption of a variety of notification systems, with unique approaches to be developed for individual schools.[23] Following publication of the report, AG Cooper worked with North Carolina legislators to pass a new law requiring court clerks to enter mental health commitments into a national gun permit database.[24]

Since the Virginia Tech shootings, which left thirty-two people dead and seventeen more injured, mobile technology, invention, and applications have improved considerably. Kristina Anderson, a survivor of the Virginia Tech shootings, recently launched an application that aims to help police better prevent and respond to campus crime. LiveSafe connects students and campus police in a two-way dialogue via their smartphones.[25]

A number of schools have followed suit. The University of Chicago announced its plans to launch a smartphone app called Pathlight to allow students to opt in to GPS tracking services.[26] In 2012, the University of North Carolina at Charlotte tested a new application called the Effective Emergency Response Communication (EERC) System for iPod Touches,[27] and Northwestern State

University in Louisiana launched a Personal Guardian application that allows users to opt in to a feature that tells police where they are going and when they arrive.[28]

Since 2000, mass notification systems have evolved from disparate systems, which were often slow and clunky, into fully integrated, robust systems.[29] Universities have invested big dollars in mass notification systems, adding layers of needed redundancy, such as digital signage in classrooms and meeting areas, indoor and outdoor sirens, social media outlets such as Facebook and Twitter, computer pop-ups, and wireless alerts.[30] This reflects the increasing variety of ways in which people access information.

Florida State University's "easy button" mass notification system is an example of one of the most ambitious systems in the country.[31] That network deploys a variety of new technologies, such as reply-to-text alerts, whereby students solicit via text the location of the nearest safe exit, and cloud-hosted emergency notification systems that never go down and can be accessed from anywhere.[32] University officials caution there is still much work to be done to improve system integration in their communications networks. In some of these institutions, funding is a major barrier, particularly for community colleges, which tend to be underfunded.[33]

First Responders and Public Safety

A number of public authorities have identified interoperable communications as a top security concern. The *9/11 Commission Report* highlighted this area in its 2004 recommendations and indicated how important it was to have communications systems that could talk to one another in real time.[34] The National Governors Association Center for Best Practices has recently identified interoperable communications as the nation's top priority in the national security area.[35]

The importance of such communication was highlighted by the shootings at Columbine High School in 1999 and the terrorist attacks on the World Trade Center and the Pentagon in 2001, and reverberated during recent tragedies such as the Boston Marathon bombings in April 2013. As noted in the aforementioned campus safety report, first responders are dependent on fast, reliable communications during and after national tragedies and natural disasters.[36] While mobile technology has improved significantly since 2000, concerns following recent national tragedies have demonstrated there is room for improvement.

As the use of smartphones becomes increasingly more common, more people are turning to mobile applications for information on preparing for and responding to emergency situations than ever before. Private companies and netizens alike are creating new applications for first responders and affected citizens. In 2010, IBM patented a natural disaster warning system that gathers data provided by MEMS accelerometers, or vibration sensors, and analyzes the information generated by seismic events.[37]

Verizon Wireless provided wireless technology— including XORA, a field-force management application that tracks SWAT vehicles; LiveCast, a live video streaming application; Blue Force, a hand-propelled, video-equipped drone and K-9 tracker; and FuzeBox, another videoconferencing application—for Alameda County Urban Shield 2011, a Department of Homeland Security exercise for first responders.[38]

The Mobile Emergency Alert System (M-EAS), demonstrated at the Association of Public-Safety Communications Officials' 2012 annual conference, uses mobile digital TV broadcasting to provide media alerts. Because the system relies on digital broadcasting instead of wireless networks, M-EAS provides content to multiple users simultaneously. Using LG mobile devices, M-EAS features both audio and visual emergency alerts sent to mobile phones, tablets, and the APCO-25 standardized emergency responder radios.[39]

Raytheon has introduced an application to improve network reliability during emergencies. Its One Force mobile application, launched in August 2012, allows first responders to communicate via data networks when outside the standard land mobile radio coverage. These "virtual radios," modeled after commercial and military designs, are reliable and secure, and provide real-time communication. In addition to voice capabilities, the One Force mobile application is equipped with maps, drawing tools, real-time positioning, video streaming, and image sharing.[40]

Motorola's PremierOne application for first responders also accesses and shares information across multiple applications and platforms, such as real-time information from computer-aided dispatch systems, records management systems, and local, state, and federal databases.[41] Both of these applications focus on facilitating communication between first responders to improve response time and accuracy.

Other first responder applications include InterAct, a data source used to dispatch instructions and provide driving directions, incident details, and responder GSP locations to law enforcement officials, and Ping4Alerts!, which allows community subscribers to receive instant notifications from law enforcement or public safety officials via mobile devices with cell, wifi, or GPS connectivity. These applications are critical for public safety officials, who rely on fast and accurate communications.

Another application, NowForce, transforms mobile devices into "lifesaving networks" through the use of geographic information systems (GIS) and has provided significant benefits to first responders in developing nations. In a country such as Nigeria, whose poor infrastructure impedes transportation and contributes to a poor electrical supply, law enforcement command and dispatch is incredibly difficult, often severely hindering reaction time. A month after local police departments launched NowForce, heavily armed response time fell to seven minutes in a city of 6 million people with heavy traffic to match.[42]

Public safety applications have proliferated in recent years, largely in reaction to specific incidents, such as the startling rise in reported crimes against women in India. According to India's National Crime Records Bureau, reported crimes against women rose 6.8 percent between 2011 and 2012 and 24.7 percent between 2008 and 2012, reflecting rising concern regarding violence against women and girls in that country.[43]

In 2013, the National Association of Software and Services Companies (NASSCOM) held an All India App Fame Contest, focused on developing solutions for female safety. Submissions included GoSuraksheit, by Hughes Systique India, an application that allows users to seek help from their most trusted contacts from any location; Sentinel, by MindHelix Technosol PVT, a GPS-based mobile application that allows users to send their GPS location to the company's security division, friends, or police officers through SMS alerts; and Nirbhaya: Be Fearless, by SmartCloud Infotech, a very accurate GPS application that sends SOS messages with the user's location to predefined SMS contact groups.[44]

In El Salvador, mobile technology is enabling real-time monitoring and analysis of crime patterns to improve crime prevention. The Santa Tecla Municipality, the U.S. government through its Agency for International Development (USAID), RTI International (RTI), the National Civilian Police (PNC), and Qualcomm Wireless Reach provide law enforcement personnel with mobile and web-based applications that allow municipalities to map and analyze rcrime data in real time. Law enforcement officers use the smartphone application, built-in camera, GPS capability, and other features to create detailed reports on crime incidents. Once a report is completed, the officer can immediately send it from the smartphone to an Internet crime database.[45]

In the wake of the surge in rape and gender-based violence cases, the Indian IT provider Tech Mahindra provided its mobile application, Fightback, for free. The application enables real-time tracking and sends alerts to five preselected contacts. Tech Mahindra is currently working with the Delhi police to integrate the application into their systems.[46]

Around the world, public safety applications are calling on citizens to use mobile technology to inform law enforcement agents of crimes and accidents. Taiwan's Ministry of the Interior, for example, launched a mobile application that allows users to access safety service and help the police fight crime. It features the 110 police service line, 113 abuse prevention line, and 165 fraud prevention line, in addition to fugitive and stolen car information, missing persons reports, road traffic updates, and taxi calling services.[47]

In Africa, law enforcement officials are taking advantage of the boom in mobile devices to educate and engage citizens. Kenyan law enforcement officers and humanitarian workers have launched a phone application comprising interactive games users can play with their friends that simulates first aid exercises designed to supplement training programs run through the Kenyan Red Cross. No real emergency services for those injured in traffic accidents currently exist in Kenya, and as motorbikes become increasingly popular, bike accidents are on the rise. The idea behind the application is to train riders—who often reach the site of an accident before first responders—in basic emergency aid to administer at the scene. This has the potential to save hundreds of lives.[48]

Law enforcement officers are working together across borders to better utilize new technology. In 2013, the U.S. Federal Communications Commission and its Canadian counterpart, Industry Canada, reached an agreement to share spectrum across their common border areas. As part of the agreement, the countries will share spectrum in the 3,650–3,700 MHz band, allowing public safety licensees on both side of the border to implement wireless broadband and high-speed Internet services.[49] The agreement is designed to improve public safety communications by facilitating spectrum sharing without the harmful interference of other bandwidths.[50]

Using Mobile Technology to Overcome Challenges

Japan's tsunami made it clear to the nation's leaders that the country needed communications networks that were, in the words of

Japanese Information Technology officials, "robust, resilient and dependable." Japanese inventors looked to Taiwan's SMS open standard, called Open GeoSMS, as a model. The tool, approved by the Open Geospatial Consortium, has been very effective in terms of humanitarian coordination and disaster relief.[51]

In recent years, Japan has combined cellular, regional wifi, and satellite networks to create a disaster-resilient, multilayered communications network.[52] 3G cell phones equipped to receive earthquake early warning systems (EEW) have been on the market in Japan since 2007. And three major Japanese mobile carriers, NTT docomo, au, and SoftBank mobile, have launched simultaneous broadcast systems to receive these alerts.[53]

When landlines go down in storms, mobile networks often stay up, enabling social networking and mobile applications to keep the lines of communication open for emergency crews, first responders, and citizens.[54] Emergency services that are using Twitter successfully can inform the public and source information from Twitter feeds during emergencies. Taiwan's 2009 Morakot Typhoon demonstrated the usefulness social media can have during disasters: web users instantly updated PTT, a popular Taiwanese social network, with information regarding the typhoon—even before the government and mass media[55] Twitter also played a critical role in the 2011 floods in Queensland, Australia; the 2011 Arab Spring uprisings; and the 2011 earthquake and tsunamis in Japan.[56] Today, Japan's Internet disaster message boards allow users to publish messages regarding their safety as a means to communicate with concerned family and friends.[57]

In the areas of disaster response and recovery, emergency technicians and hospitals are critical. A recent Italian pilot study tested an emergency system solution called Mobile Emergency that is designed to improve the readiness of hospital personnel during disasters and facilitate the treatment of and procedures on victims of disasters. The application stores hospital map information, sends emergency alarms, provides personnel support during the disaster, and allows users to locate nearby colleagues.

The application was piloted at the Careggi Hospital in the Florence and Tuscany area in Italy and yielded encouraging results. The pilot found that the total time from receiving an emergency call to reaching the emergency scene dropped from an average of sixteen minutes and twenty seconds to nine minutes and fifty-seven seconds, a 35 percent reduction.[58] These results demonstrate the ways in which mobile technology is improving speed and accuracy for medical professionals and first responders. The use of mobile technology at every stage of disaster, from preparation to response to recovery, has the very real potential to reduce errors and save lives.

An illustration of an inexpensive application for firefighters in Latin America further demonstrates the usefulness of mobile technology. In many parts of Latin America, firefighters are volunteers who receive minimal support or funding from government agencies. Applications such as MobileMap largely reduce the need for costly equipment. This mobile tool runs on handheld devices such as cell phones, laptops, and notebooks. The feature allows firefighters in Latin America to communicate with central command centers through wifi, cellular, or, if these are unavailable, a mobile ad hoc network (MANET). Through preloaded maps and more accurate and robust communication features (such as image file transfer), MobileMap greatly improved firefighter arrival time and accuracy during pilot tests in Santiago, Chile.[59]

Providing reliable facts and up-to-date information through social media outlets and encouraging citizens to tweet, text, or blog crucial information to first responders and government officials is an easy and effective way to coordinate responses. For example, the Philippines used Twitter to update and engage its citizens during disaster. During its 2012 typhoon, the Philippine government suggested hashtags for followers to provide important information to officials and stay updated on the evolving situation.[60]

Hong Kong and Australia turned to mobile technology to mitigate rumors during periods of crisis. At the height of the SARS pandemic, the government of Hong Kong sent SMS messages to 6 million mobile phones in an attempt to allay fears surrounding supposed government action.[61] These messages directed people via 3G networks to websites that provided maps, videos, and other kinds of helpful information. In Australia, Queensland police created a "mythbusters" hashtag to manage misconceptions and diffuse rumors during the floods.[62]

As with any form of instant communication, information verification is a top priority. During the April 2013 Boston Marathon bombing, some untrue statements permeated the social media, including the suggestion that the government had shut down cell phone networks to prevent remote bomb detonation. But, in fact, the Boston Police Department used mobile social media to provide relevant and timely information. Its Twitter feed increased from approximately 35,000 followers to nearly a quarter of a million, thereby revealing the desire for trusted and reliable information during a time of great public concern.[63]

Maintaining a functioning system is another challenge. After the earthquake in Haiti in 2010, conventional phone networks were knocked offline for forty-eight hours.[64] In the days after Hurricane Sandy hit the northeastern United States in October 2012, nearly a quarter of the cell phone towers in ten East Coast states were damaged or destroyed. Data centers submerged in water shut down websites such as *Buzzfeed, Gawker, Gizmodo,* and the *Huffington Post* and briefly affected traders on the floor of the New York Stock Exchange.[65]

Broadband networks require adequate bandwidth and reliable sources of electrical power. Some cell towers are tied to the electric grid, making them vulnerable to strong winds, flooding, or natural disasters. While many cell towers are equipped with backup generators, some of them operate only for limited durations of a day or so.[66] We need resilient networks with sufficient spectrum that can operate during lengthy power outages or massive call volume spikes.[67]

There has been some progress on this front, such as building redundant networks where necessary, utilizing portable or temporary base stations during emergencies, relying on backup power sources for cell stations and switches during electrical grid failures, and tailoring network needs to individual locations and disasters.[68] In addition, organizations such as the International Telecommunications Union (ITU) have provided assistance, for example during the 2010 Haiti earthquake, when the ITU deployed forty satellite terminals as well as a deployable base station to reestablish basic communication.[69]

As people increasingly rely on mobile phones and the Internet as their main conduit and source of information and communication, the government and private sector must work together to ensure that systems respond effectively to natural disasters and have the bandwidth to maintain operability during times of massive volume increases.

Many police departments and first responders currently have public safety systems that are not interoperable. This makes it difficult for them to communicate with each other during times of major disasters or public safety challenges. They often end up relying on commercial networks owing to their ubiquity and reliability.[70]

In future efforts to improve interoperability, it is important to build on technologies that connect users. For example, LTE-direct and small cell technology make it possible for emergency responders to stay in touch during periods of crisis. They provide low-power, autonomous access to users across distances of up to one kilometer. These and other kinds of device-to-device communications can help people stay in touch during local emergencies.

The Department of Commerce's National Telecommunications and Information Administration (NTIA) is working on the creation of a nationwide interoperable wireless broadband network for public safety officials. This First Responder Network Authority, or FirstNet, is an independent authority within NTIA and is assigned to building, deploying, and operating the FirstNet network. [71]

Such efforts should take advantage of commercial applications that directly connect users and offer network redundancy. In emergency situations, it is valuable to have multiple modes of connection. If one system is compromised, backup systems will then be available to help connect emergency personnel.

There is no way to prevent future natural disasters. It is possible, however, to make better use of the current mobile technologies and promote future advances. As mobile devices become more common around the globe, policymakers have the chance to provide first responders and citizens with the tools necessary to save lives during threatening events. Governments and businesses should take full advantage of these technologies in order to help people and mitigate suffering.

10 FUTURE PROSPECTS

THE MOBILE REVOLUTION IS in full swing. Innovative, mobile technology–based solutions in education, health care, entrepreneurship, poverty alleviation, public outreach, and disaster relief, among other areas, are transforming those sectors. In both developed and developing economies, people, businesses, and governments are harnessing the power of wireless technology for useful purposes.

Yet despite considerable progress, the mobile world remains in its infancy. Some believe that in shifting from desktop computing to mobile devices, people are simply changing platforms while holding everything else the same. Those who hew to this notion underestimate the fundamental nature of the transformation and how mobile technology alters operations at many different levels. They do not grasp that optimizing services for mobile technology involves changing organizations, operations, and culture. Conversely, merely dropping mobile devices into traditional enterprises without making any other changes limits the possibilities and makes it difficult to achieve the full advantages of the revolution. Going mobile means that leaders have to retool their entire operation for a wireless world.

In this chapter, I look ahead and analyze the challenges that remain. Important issues include the emerging "Internet of Things," prospects for real-time decisionmaking, boosting entrepreneurship, growing the economy, strengthening government performance, lowering barriers to innovation, and extending mobile infrastructure. Making progress in these areas would help spread the benefits of mobile technology more broadly around the world.

The Internet of Things

Mobile connectivity is very important for what has come to be known as the "Internet of Things." In 2009, the writer Kevin Ashton coined that phrase to describe the emergence of machine-to-machine communications linked through high-speed networks and cloud-based solutions.[1] Five years later, what started as a theoretical construct has emerged full-blown in practice and is changing people's lives in fundamental ways.

Smart objects enable manufacturers to track their supply chains more effectively. Through digital tags attached to particular items, managers can see where their supplies are and whether they have the inventories needed to build particular products. This type of connectivity used in conjunction with global positioning systems (GPS) transforms different areas. The consultant Nam Pham has analyzed the situation and concluded that "with the increased adoption of mobile phones throughout the world and the growth of phones with GPS positioning capabilities, mobile phones and networks are now an essential tool for things like agriculture, transportation/logistics and emergency response and disaster management, as well as providing important tools and information to protect and aid individuals."[2]

Smart appliances help consumers keep track of their heating and security needs. For example, consumers can use mobile devices to set their thermostats or turn on their home security systems. They can make sure that dishwashers run at optimal times from the standpoint of saving energy and that refrigerators keep products cool. Smartmeters show people how they are consuming energy and what the cost is at various points in time.

Motor vehicles now are equipped with GPS chips that monitor engine performance and ensure that the cars are operating at peak efficiency. Smart cars can help people park in tight spaces and anticipate possible accidents through early warning systems.

Wearable devices are becoming more common. People are wearing computerized watches, glasses, and accessories that keep them connected around the clock. Innovators are integrating devices into people's lives rather than developing computers that function as separate devices.

As the world evolves from human-to-human computing to sensors and machine-to-machine communications, mobile devices will have even greater value than today. Once they are embedded in a range of daily activities, it will become apparent how much they contribute to a number of different endeavors. According to the wireless market research and analysis firm Maravedis, "20% of mobile carriers say M2M [machine-to-machine] will be a number one source of revenue growth 2013–2017."[3]

Prospects for Real-Time Decisionmaking

Mobile networks enable real-time decisionmaking and data analysis. In areas from transportation and energy to education to health care, leaders are figuring out how to mine information for useful trends and how to use these data to improve decisionmaking. Rather than spending weeks or months collecting data and then converting them to usable form, it is possible to use social media information, search data, consumer complaints, traffic patterns, health care choices, and learning data to identify both trouble spots and opportunities for improvement.

It is not easy to compile and analyze data. There are interoperability challenges arising from hardware devices and software that don't connect smoothly. Data often come in forms that are not compatible or that do not have the same meanings in different contexts. Increasingly, there are unstructured data that are not easily tabulated. Consumers worry about privacy protection and maintaining security in their information networks.

But if these difficulties can be overcome, it is possible to help leaders make more informed choices and to reduce the information gap between leaders and citizens. Right now, people often feel isolated from those who make decisions on their behalf. Improving connectivity and interactivity are ways to facilitate positive change around the world.

Boosting Entrepreneurship

The mobile economy benefits from entrepreneurship. Many of the inventors who played a crucial role in the development of mobile devices and services started out working by themselves— the garage inventors—or in small organizations. Despite the small size of their operations, they came up with advances in hardware and software that propelled the mobile ecosystem into a global industry that generates more than $1.6 trillion in revenue.[4]

Yet governments vary enormously in the amount they invest to support R&D and how seriously they take invention and creativity. Some countries devote considerable resources to encouraging entrepreneurship and launching new firms, while others do not see that as an important priority. But regardless of the particular area, new companies create jobs and spur economic growth.

Entrepreneurship is crucial for economic development. Mobile devices help innovative individuals access capital, locate market information, and reach new consumers; they also empower women and less advantaged groups.[5] According to a World Bank analysis of ways to maximize mobile use, software applications have become one of the greatest areas for entrepreneurship and economic growth.[6]

Growing the Economy

One of the reasons many countries invest in mobile infrastructure and encourage mobile use is the relationship of wireless technology to economic growth. An analysis by the economists Harald Gruber and Pantelis Koutroumpis, for example, found that national growth improves significantly based on mobile use. Looking at 192 nations from 1990 to 2007, they found increasing returns in terms of productivity and growth based on the use of mobile devices. For high-income nations, mobile technology added 0.20 percent annually to GDP, while in low-income countries it added 0.11 percent.[7] They also looked at mobile infrastructure investment and found that it paid off in economic growth. Nations that invested saw GDP gains of 0.39 percent in high-income countries and 0.19 percent in low-income countries.[8] These gains underscore the value of the mobile economy.

A number of studies have looked at the economic impact of wireless-based connectivity for specific countries and have found a significant tie. For example, the consulting firm LECG in an analysis of mobile technology in India projected that "an investment of US$20 billion in 3G networks over the next five years will benefit India's economy by more than US$70 billion and create up to 14 million jobs." For China, the same firm projected that "the current and future economic benefits from Chinese operators' proposed investment of $59 billion is likely to exceed $110 billion."[9]

This finding was echoed by an Analysis Mason project, which projected that an "increase in broadband penetration of 1% will contribute INR 162 bn, or 0.11% to Indian GDP in 2015." The report also indicated that allocating an "additional 5 MHz of 3G spectrum will increase BB penetration by 3.3% of the population and enhance GDP by INR 538 bn in 2015."[10]

In examining the European market in general, the GSM Association found that mobile investment was key to higher economic growth. Its report proposed that investment in mobile technology "could generate approximately €60bn to €120bn in value annually to 2015—equivalent to 0.5% to 1.0% of GDP—or €340bn–€750bn in aggregate between 2010 and 2015."[11]

For the United States, a Deloitte analysis estimated that small and medium-size enterprises with Internet access experienced an average 11 percent productivity gain, and that extending Internet access to current levels in developed countries could enhance productivity by as much as 25 percent in developing economies. The consulting firm calculated that the resulting economic activity would generate $2.2 trillion in additional GDP, a 72 percent increase in the GDP growth rate, and more than 140 million new jobs.[12]

Similar trends are apparent in the developing world. GSM research indicates that the mobile economy is likely to add $729 billion to Asian Pacific GDP by 2020. It is anticipated that 2 million new jobs will be created and there will be $131 billion in new tax revenues. In the case of India, analysts have found that "states with 10 per cent higher than average mobile phone penetration enjoy an annual average growth rate 1.2 per cent higher than those with a lower teledensity." This observation led researcher Rajat Kathuria of the Indian Council for Research and International Economic Relations (ICRIER) to conclude that "telecommunications is a critical building block for the country's economic development. Our work also shows that the real benefits of telecommunications only start when a region passes a threshold penetration rate of about 25 percent. Many areas have still not attained that level, which indicates the importance of increasing teledensity as soon as possible."[13]

Research by Facebook and Internet.org suggests that if there were universal Internet access around the world, it would connect the two-thirds (or 5 billion) of the world's population without online connections and would "increase productivity by as much as 25 percent, generating $2.2 trillion in GDP and more than 140 million new jobs, lifting 160 million people out of poverty."[14]

Strengthening Government Performance

There is evidence that mobile technology helps to improve government performance. By increasing transparency and accountability, there are positive impacts on political and economic systems. Several countries are devoting considerable efforts to using technology to improve transparency and accountability and encouraging people to get more involved in the political process.

For example, the Technology for Transparency Network relies on social media and mobile technology to fight corruption. According to its website, "In Kenya, Mzalendo seeks to make information more accessible from the proceedings of the country's parliament. In Jordan, Ishki aims to involve citizens in developing solutions to civic problems. Vota Inteligente in Chile promotes government transparency by informing Chilean citizens about corruption and policy debates through the use of social media."[15]

The Need to Extend the Mobile Infrastructure

With the growing importance of mobile devices, one of the most important issues today is building the infrastructure that enables the mobile economy. There are challenges on several fronts, including expanding high-speed mobile networks, improving access to underserved populations, and encouraging an environment that supports invention and creativity.

Having fast mobile networks is crucial for the utilization of m-health, m-learning, smart appliances, and smartmeters. Taking advantage of the opportunities provided by mobile technology requires networks that can handle those kinds of applications. Some places lack sufficient spectrum to support the growing use of wireless devices. For example, a 2011 Credit Suisse report estimated that North American mobile networks are nearing capacity in certain locales owing to dramatic increases in video streaming and other bandwidth-intensive applications. With Internet traffic rising rapidly, it is vital that investors build out the networks that support the "Internet of Things."[16]

This conclusion is not limited to the developed world but is also relevant for developing nations. A study of m-commerce in China found that it is growing in importance. Using survey data from China's Ministry of Commerce, researchers in 2008 found that "m-commerce in China enjoys relatively great communication reach" and that the "costs of communications do not seem to be a major barrier for m-commerce growth."[17]

It also is vital to improve access to underserved populations. It will be hard to realize the promise of the mobile economy unless all have the chance to share in its benefits. People in rural areas often do not have access to mobile networks, and this makes it difficult for them to access m-health or m-learning platforms. They are not able to reap the benefits of the mobile revolution if they face connectivity problems in terms of infrastructure or device usage.

The mobile revolution has spread because of advances in microchips and cellular communications. Invention is crucial for the continued growth of the mobile economy. There are many aspects to fostering an environment that makes this possible, and the countries that have dedicated resources and encouraged activity as it relates to securing and valuing intellectual property will benefit the most. Having a vibrant mobile ecosystem depends on the continued ingenuity of designers and manufacturers. Only by developing faster and more efficient means of handling mobile transactions and accessing information can the global economy continue to thrive.

NOTES

Chapter 1

1. Portions of this chapter are drawn from Darrell West, "Ten Facts about Mobile Broadband," Brookings Policy Report (Brookings Institution, December 8, 2011).

2. A. T. Kearney, "GSMA: The Mobile Economy" (London, 2013).

3. Ibid., p. 11.

4. Ibid., p. 5.

5. Darrell West, "How Mobile Technology Is Driving Global Entrepreneurship," Brookings Policy Report (Brookings Institution, October 23, 2012).

6. Jenny Aker and Isaac Mbiti, "Mobile Phones and Economic Development in Africa," Working Paper 211 (Washington: Center for Global Development, June 2010).

7. Ken Hyers, "A Peek into the Future of Mobile" (Boston: Strategy Analytics, January 2011).

8. Pew Research Center, "Mobile Technology Fact Sheet" (Washington, December 27, 2013).

9. Ibid.

10. Deloitte, "The Impact of 4G Technology on Commercial Interactions, Economic Growth, and U.S. Competitiveness" (Chicago, August 2011) (www.deloitte.com/assets/Dcom-UnitedStates/Local%20Assets/Documents/TMT_us_tmt/us_tmt_impactof4g_081911.pdf), p. 1.

11. Aker and Mbiti, "Mobile Phones and Economic Development in Africa."

12. Strategic Networks Group, "Economic Impact Study of the South Dundas Township Fibre Network" (London: United Kingdom Department of Trade and Industry, July 27, 2003).

13. Raul L. Katz, Stephan Vaterlaus, Patrick Zenhausern, Stephen Suter, and Philippe Mahler, "The Impact of Broadband on Jobs and the German Economy," unpublished paper, Columbia Business School, New York, 2009.

14. Christine Zhen-Wei Qiang, "Telecommunications and Economic Growth," unpublished paper, World Bank, Washington, 2009.

15. Darrell West, *Digital Schools: How Technology Can Transform Education* (Brookings Institution Press, 2012).

16. Mark Schneiderman's comments quoted in "Innovate to Educate: System [Re]Design for Personalized Learning" (Washington: Software & Information Industry Association, 2010), p. 8.

17. Ruth Moody and Michael Bobic, "Teaching the Net Generation without Leaving the Rest of Us Behind: How Technology in the Classroom Influences Student Composition," *Politics & Policy* 39, no. 2 (2011): 169–94.

18. Jessica Briskin, Tutaleni Asino, Michael Montalto-Rook, and Yaozu Dong, "Smart Apps: An Analysis of Educational Applications Available on Smartphones and the Implications for Mobile Learning," paper presented at the annual convention of the Association for Educational Communications and Technology, Anaheim, Calif., 2010.

19. Stephoni Case and Susan Stansberry, "Teaching with Facebook as a Learning Management System," paper presented at the annual convention of the Association for Educational Communications and Technology, Anaheim, Calif., 2010.

20. Chris Dede, "Immersive Interfaces for Engagement and Learning," *Science* 323 (January 2, 2009): 66–69.

21. C. Shuler, "Pockets of Potential: Using Mobile Technologies to Promote Children's Learning" (New York: Joan Ganz Cooney Center at Sesame Workshop, 2009).

22. M. Lu, "Effectiveness of Vocabulary Learning Via Mobile Phone," *Journal of Computer Assisted Learning* 24, no. 6 (2008): 515–25.

23. Minjuan Wang, Riumin Shen, Daniel Novak, and Xiaoyan Pan, "The Impact of Mobile Learning on Students' Learning Behaviors and Performance," *British Journal of Educational Technology* 40, no. 4 (2009): 673–95.

24. Project Tomorrow, "From Chalkboard to Tablets: The Digital Conversion of the K–12 Classroom," a Speak Up survey, Tomorrow.

org, April 2013 (www.tomorrow.org/speakup/SU12_DigitalConversion_EducatorsReport.html).

25. Darrell West, "Customer-Driven Medicine: How To Create a New Health Care System" (Brookings Institution, October 2009).

26. Mirela Prgomet, Andrew Georgious, and Johanna Westbrook, "The Impact of Mobile Handheld Technology on Hospital Physicians' Work Practices and Patient Care," *Journal of the American Medical Informatics Association* 16 (November/December 2009): 792–801.

27. Chris Sweeney, "How Text Messages Could Change Global Healthcare," *Popular Mechanics,* October 26, 2011.

28. Adam Thierer, "Creating Local Online Hubs," White paper (Washington: Aspen Institute, 2011).

29. Darrell West, "An International Look at High-Speed Broadband" (Brookings Institution, December 2009).

30. Renate Steinmann, Alenka Krek, and Thomas Blaschke, "Analysis of Online Public Participatory GIS Applications with Respect to the Differences between the US and Europe" (Salzburg Research Forschungsgesellschaft, University of Salzburg, Austria, 2004).

31. G. S. Hanssen, "E-communication: Strengthening the Ties between Councillors and Citizens in Norwegian Local Government," *Scandinavian Political Studies* 31, no. 3 (2008): 333–61.

32. H. Rojas and E. Puig-i-Abril, "Mobilizers Mobilized: Information, Expression, Mobilization and Participation in the Digital Age," *Journal of Computer Mediated Communication* 14, no. 4 (2009): 902–27.

Chapter 2

1. Cherie Blair Foundation for Women, Exxon Mobil, and Booz & Co., "Mobile Value Added Services: A Business Growth Opportunity for Women Entrepreneurs" (London, 2012), p. 8.

2. David Neumark, Brandon Wall, and Junfu Zhang, "Do Small Businesses Create More Jobs?," *Review of Economics and Statistics* 93, no. 1 (August 2011): 16–29.

3. Maja Andjelkovic, "The Future Is Mobile: Why Developing Country Entrepreneurs Can Drive Internet Innovation," *SAIS Review* 30, no. 2 (Summer–Fall 2010).

4. United Nations Conference on Trade and Development (UNCTAD), "Information Economy Report 2011" (New York and Geneva, 2011) (http://unctad.org/en/PublicationsLibrary/ier2011_en.pdf), p. xiii.

5. Vodafone Group, "Africa: The Impact of Mobile Phones: Moving the Debate Forward," Vodafone Policy Paper Series 2 (Newbury, Berkshire, March 2005).

6. Portions of this chapter are drawn from Darrell West, "How Mobile Technology Is Driving Global Entrepreneurship," Brookings Policy Report (Brookings Institution, October 23, 2012).

7. "How Has Wireless Technology Changed How You Live Your Life?," *Time*, August 27, 2012, pp. 34–39. The *Time* Mobility Poll was undertaken in cooperation with Qualcomm between June 29 and July 28, 2012. The margin of error for the survey as a whole is plus or minus 1.5 percentage points.

8. UNCTAD, "Information Economy Report 2011," p. 19.

9. Calestous Juma, "Africa's New Engine," *Finance and Development* 48, no. 4 (December 2011): 6–7.

10. International Center for Research on Women, "Connectivity: How Mobile Phones, Computers and the Internet can Catalyze Women's Entrepreneurship. India: A Case Study" (New Delhi, India, 2012), (http://www.icrw.org/files/publications/Connectivity-how-mobile-phones-computers-and-the-internet-can-catalyze-womens-entrepreneurship.pdf), p. 14.

11. Jenny Aker and Isaac Mbiti, "Mobile Phones and Economic Development in Africa," *Journal of Economic Perspectives* 24, no. 3 (Summer 2010): 207.

12. Qualcomm Wireless Reach, "Mobile Microfranchise and Application Laboratory Initiatives Give Entrepreneurs New Tools for Success" (San Diego, June 2012).

13. Ibid.

14. Karishma Vaswani, "Indonesian Farmers Reaping Social Media Rewards" (London: British Broadcasting Corporation, June 4, 2012).

15. Aker and Mbiti, "Mobile Phones and Economic Development in Africa," p. 217.

16. Ibid., p. 218.

17. International Telecommunications Union, *World Telecommunication/ICT Indicators Database,* June 2014.

18. Qualcomm Wireless Reach, "Internet Cafes: Creating a Communications Gateway to the Unconnected" (San Diego, July 2012).

19. Qualcomm Wireless Reach, "Fishing with 3G Nets: Promoting Sustainable Fishing and Entrepreneurship through Digital & Social Inclusion" (San Diego, July 2012).

20. Kim Mallalieu and Mark Lessey, "Mobile Apps Boost Trinidad and Tobago Fish Market," *Digital Opportunity,* March 21, 2012.

21. Qualcomm Wireless Reach, "Internet Cafes."

22. World Association of Newspapers and News Publishers, "Mobile Media Services at Sub-Saharan African Newspapers" (Nairobi, Kenya, July 2011), pp. 30–31.

23. Robert Jensen, "The Digital Provide: Information (Technology), Market Performance, and Welfare in the South Indian Fisheries Sector," *Quarterly Journal of Economics* 122, no. 3 (August 2007): 883.

24. GSMA Development Fund and the Cherie Blair Foundation for Women, "Women & Mobile: A Global Opportunity" (Mountain View, Calif.: Vital Wave Consulting, 2013) (www.gsma.com/mobilefor development/wp-content/uploads/2013/01/GSMA_Women_and_Mobile-A_Global_Opportunity.pdf), p. 39.

25. UNCTAD, "Information Economy Report 2011," p. 75.

26. Martin Carstens, "30 Brilliant Africa Tech Startups," *Venture Burn,* March 9, 2012.

27. Cherie Blair Foundation for Women and others, "Mobile Value Added Services," p. 12.

28. GSMA Development Fund and the Cherie Blair Foundation for Women, "Women & Mobile," p. 6.

29. Ibid., p. 3.

30. Qualcomm Wireless Reach, "Global Ready eTraining Centers: Technology Training and Job Opportunities for Underserved Indonesians" (San Diego, July 2012).

31. International Center for Research on Women, "Connectivity," p. 38.

32. UNCTAD, "Information Economy Report 2011," p. 21.

33. Aker and Mbiti, "Mobile Phones and Economic Development in Africa," p. 221.

34. Claudia McKay and Mark Pickens, "Branchless Banking 2010: Who's Served? At What Price? What's Next?," Consultative Group to Assist the Poor Report 66 (Washington, September 1, 2010).

35. World Association of Newspapers and News Publishers, "Mobile Media Services at Sub-Saharan African Newspapers," p. 18.

36. GSMA Development Fund and the Cheri Blair Foundation for Women, "Women & Mobile," p. 29.

37. Cherie Blair Foundation for Women and others, "Mobile Value Added Services," p. 17.

38. UNCTAD, "Information Economy Report 2011," p. 17.

Chapter 3

1. World Bank data available at http://povertydata.worldbank.org/poverty/home/.

2. Portions of this chapter are drawn from Darrell West, "Alleviating Poverty: Mobile Communications, Microfinance and Small Business Development around the World," Brookings Policy Report (Brookings Institution, May 16, 2013).

3. Jeffrey Sachs quoted in Kyla Yeoman, "Can Mobile Phones End Extreme Poverty? Jeffrey Sachs Thinks So," GlobalEnvision.org, March 16, 2012 (www.globalenvision.org/2012/03/16/can-mobile-phones-end-extreme-poverty-jeffrey-sachs-thinks-so).

4. "How Has Wireless Technology Changed How You Live Your Life?," *Time*, August 27, 2012, pp. 34–39. The *Time* Mobility Poll was undertaken in cooperation with Qualcomm between June 29 and July 28, 2012.

5. World Economic Forum, "Global Agenda Session on Anti-Corruption" (Davos, 2012). See also Transparency International, "Corruption Perceptions Index 2011" (Berlin, 2011) (www.transparency.org/cpi2011/results).

6. United Nations Global Compact, "Fighting Corruption in the Supply Chain," UNGlobalCompact.org, June 2010 (http://supply-chain.unglobalcompact.org/site/article/17).

7. Mehmet Ugur and Nandini Dasgupta, "Evidence on the Economic Growth Impacts of Corruption in Low-Income Countries and Beyond" (London: EPPI-Centre, University of London, 2011), p. 2.

8. Daniel Bond, Daniel Platz, and Magnus Magnusson, "Financing Small-Scale Infrastructure Investments in Developing Countries," DESA (Department of Economic and Social Affairs) Working Paper 114, UN.org, May 2012 (www.un.org/esa/desa/papers/2012/wp114_2012.pdf), p. 1.

9. Jones Lang LaSalle, "Real Estate Transparency Back on Track," Global Foresight Series 2012, LaSalle.com (http://www.lasalle.com/Research/ResearchPublications/TransparencyIndex_2012.pdf).

10. Transparency International, "Corruption Perceptions Index 2011."

11. McKinsey & Company, "Global Financial Inclusion" (New York, Fall 2010) (www.thegiin.org/binary-data/RESOURCE/download_file/000/000/149-1.pdf).

12. Ibid.

13. See the website www.eko.co.in.

14. Vishy, "Empowering Bihar's Rural Health Workers with Mobile Phones and Money Transfer," TechSangam.com, September 8, 2011 (www.techsangam.com/2011/09/08/empowering-bihar-rural-health-workers-with-mobile-phones-and-money-transfer/).

15. Jennifer Schenker, "MPedigree's Tx for Counterfeit Drugs," *Business Week*, December 3, 2008.

16. Bright Simons, "Medigree—An African Blueprint for Consumer Empowerment," *All Africa*, January 15, 2013 (http://allafrica.com/stories/201301150815.html).

17. Description shown at http://farmerline.org.

18. *Farmerline Blog*, "Women in Aquaculture Embrace Farmerline," February 18, 2013.

19. Brad McCarty, "Bizness Apps Reach 1,000 Small Business Customers in 9 Months," *The Next Web*, July 1, 2011 (http://thenextweb.com/insider/2011/07/01/bizness-apps-reaches-1000-small-business-customers-in-9-months/).

20. Lamar Morgan, "GoMobiMe Helps Small Businesses Get Found on Mobile Devices," *Examiner*, February 16, 2013.

21. KIVA, "Loans That Change Lives," April 8, 2013 (http://www.kiva.org/).

22. See the Consultative Group to Assist the Poor website at www.cgap.org.

23. See ibid.

24. See ibid.

25. Stephanie Strom, "Web Sites Shine Light on Petty Bribery Worldwide," *New York Times*, March 6, 2012.

26. Jenny Aker and Isaac Mbiti, "Mobile Phones and Economic Development in Africa," *Journal of Economic Perspectives* 24, no. 3 (Summer 2010): 207.

27. GSMA Development Fund and the Cherie Blair Foundation for Women, "Women & Mobile: A Global Opportunity" (Mountain View, Calif.: Vital Wave Consulting, 2013) (www.gsma.com/mobilefordevelopment/wp-content/uploads/2013/01/GSMA_Women_and_Mobile-A_Global_Opportunity.pdf).

28. Qualcomm Wireless Reach, "Mentoring Women in Business: Using 3G Tablets to Train Women Entrepreneurs" (San Diego, October 2012) (www.qualcomm.com/media/documents/wireless-reach-case-study-malaysia-mentoring-women-business-english).

Chapter 4

1. George Leard, "Biography: Martin Cooper," Helium.com, March 11, 2010 (www.helium.com/items/1235159-martin-cooper-mobile-phone-inventions-whartonmotorola-attwireless-technology-innovatators).

2. Portions of this chapter are drawn from Darrell West, "Invention and the Mobile Economy," Brookings Policy Report (Brookings Institution, March 5, 2013).

3. Jon Agar, *Constant Touch: A Global History of the Mobile Phone* (Cambridge: Icon, 2004).

4. "Jesse Russell" and "BusinessMakers," *The HistoryMakers: The Nation's Largest African American Video Oral History Collection,* interviewed May 16, 2012 (www.thehistorymakers.com/biography/jesse-russell).

5. See the home page of GreatCall at www.greatcall.com.

6. "Arlene Harris—Top U.S. Wireless Innovators of All Time," *Fierce Wireless* (www.fiercewireless.com/special-reports/arlene-harris #ixzz2IjyLlR00).

7. Rick Merritt, "Irwin Jacobs on Cellular's Past and Future," *EE Times,* May 5, 2011.

8. "Harold Haas: Communications Technology Innovator," *TED: Ideas Worth Spreading,* posted August 2011 (www.ted.com/speakers/harald_haas.html).

9. "Harold Haas: Communications Technology Innovator," *TED: Ideas Worth Spreading,* July 2011 (www.ted.com/talks/harald_haas_wireless_data_from_every_light_bulb).

10. Yuki Noguichi, "Young Entrepreneur Has a Better Idea. Now What?," *NPR's Morning Edition,* posted August 23, 2011 (www.npr.org/2011/08/23/139854129/young-entrepreneur-has-a-better-idea-now-what).

11. Ibid.

12. "Executive Perspectives: Maryam Rofougaran," Broadband.com.

13. Broadcom, "Broadcom's New Combo Chip Integrates 802.11n Wi-Fi, Bluetooth 4.0 + HS and FM to Bring New Multimedia Applications to Smartphones, Tables and Other Mobile Devices," press release, Barcelona, Spain, February 14, 2011 (www.broadcom.com/press/release.php?id=s549642).

14. See the website for Rearden Companies at www.rearden.com/people.

15. Ashlee Vance, "Steve Perlman's Wireless Fix," *Bloomberg Businessweek Magazine Online,* July 27, 2011 (www.businessweek.com/magazine/the-edison-of-silicon-valley-07272011.html#p1).

16. Dennis Normile and Charles Mann, "Asia Jockeys for Stem Cell Lead," *Science* 307, no. 5710 (February 4, 2005): 660–64.

17. Darrell West, *Brain Gain: Rethinking U.S. Immigration Policy* (Brookings Institution Press, 2010).

18. Michael Liedtke, "Google's Motorola Mobility Acquisition Closes," *Huffington Post,* May 22, 2012.

19. See the World Bank Patent and Intellectual Property Data Base at http://data.worldbank.org/indicator. The World Bank aggregated data from the World Intellectual Property Organization are available at www.wipo.int/ipstats/en/statistics/patents.

20. Portions of this section are drawn from Darrell West, "The State of the Mobile Economy," Brookings Policy Report (Brookings Institution, 2014).

21. Organization for Economic Cooperation and Development (OECD), Science and Technology Statistical Compendium, 2004 (Geneva, 2004) (www.oecd.org/sti/scienceandtechnologystatisticalcompendium 2004.htm).

22. National Science Board, "Science and Engineering Indicators 2008" (National Science Foundation, 2008) (www.nsf.gov/statistics/seind08/c4/c4s1.htm).

23. OECD, Science and Technology Statistical Compendium, 2004.

24. National Science Board, "Science and Engineering Indicators 2008."

25. Ibid.

26. Ibid.

27. U.S. Department of Commerce, "Intellectual Property and the U.S. Economy: Industries in Focus," March 2012 (www.uspto.gov/news/publications/IP_Report_March_2012.pdf).

28. There is a gap in the Japanese data for 1981 and 1982 in the World Bank Patent and Intellectual Property Data Base.

29. Darrell West, "Improving University Technology Transfer and Commercialization," Issues in Technology Innovation no. 20 (Brookings Institution, December 2012).

30. Moira Herbst, "Still Wanted: Foreign Talent—and Visas," *Business Week,* December 21, 2009, 76.

31. David Kappos, "How an Improved U.S. Patent and Trademark Office Can Create Jobs," statement before the Subcommittee on

Intellectual Property, U.S. House of Representatives Judiciary Committee, January 25, 2011.

Chapter 5

1. Darrell West, *Digital Schools: How Technology Can Transform Education* (Brookings Institution Press, 2012).

2. Parts of this chapter are drawn from Darrell West, "Mobile Learning: Transforming Education, Engaging Students, and Improving Outcomes," Brookings Institution Policy Report (Brookings Institution, September 17, 2013).

3. Project Tomorrow, "From Chalkboard to Tablets: The Digital Conversion of the K–12 Classroom," a Speak Up survey, Tomorrow. org, April 2013 (www.tomorrow.org/speakup/SU12_DigitalConversion_ EducatorsReport.html).

4. Susan Lund, James Manyika, Scott Nyquist, Lenny Mendonca, and Sreenivas Ramaswamy, "Game Changers: Five Opportunities for US Growth and Renewal," McKinsey Global Institute report (New York: McKinsey & Co., July 2013), p. 111.

5. Paul Kielstra, *The Learning Curve,* Economist Intelligence Unit, Pearson.com, 2012, p. 19.

6. Eric Hanushek, Paul Peterson, and Ludger Woessmann, "Achievement Growth: International and U.S. State Trends in Student Performance," Harvard Kennedy School Program on Education Policy and Governance & Education Next (Cambridge, Mass., July 2012), p. vi.

7. Tom Loveless, "How Well Are American Students Learning?" (Brookings Institution, March 2013), p. 3.

8. Ibid., p. 7.

9. Martin West, "Education and Global Competitiveness: Lessons for the U.S. from International Evidence," unpublished paper, Department of Education, Harvard University, p. 5.

10. Ibid., p. 21.

11. Richard Weingroff, "The Genie in the Bottle: The Interstate System and Urban Problems, 1939–1957," *Public Roads* 64, no. 2 (September–October 2000).

12. Federal Communications Commission (FCC), *Connecting America: National Broadband Plan* (Washington: Government Printing Office [GPO], 2010), 9–10.

13. Marguerite Reardon, "The Price of Universal Broadband," *CNET News*, September 30, 2009.

14. FCC, *National Broadband Plan*, Recommendation 11.23.

15. Project Tomorrow, "From Chalkboard to Tablets: The Emergence of the K-12 Digital Learner," a Speak Up survey, Tomorrow.org, June 2013 (www.tomorrow.org/speakup/SU12_DigitalLearners_Student Report.html).

16. NMC Horizon Report, "2013 K–12 Edition," New Media Consortium, the Consortium for School Networking, and the International Society for Technology in Education (Austin, Tex., 2013), p. 10.

17. Irwin Jacobs, "Modernizing Education and Preparing Tomorrow's Workforce through Mobile Technology", paper presented at the i4j Summit, Menlo Park, Calif., March 2013, p. 10.

18. Kathleen Fulton and Ted Britton, "STEM Teachers in Professional Learning Communities" (Washington: National Commission on Teaching and America's Future, June 2011), pp. 8–9 (http://nctaf.org/wp-content/uploads/2012/01/1098-executive-summary.pdf).

19. Becky Ham, "'Active Explorer' Project Turns Cell Phones into Research Tools for Students" (Washington: American Association for the Advancement of Science, November 13, 2012).

20. Project Tomorrow, "From Chalkboard to Tablets: The Emergence of the K–12 Digital Learner."

21. Project Tomorrow, "2013 Trends in Online Learning Virtual, Blended and Flipped Classrooms," Tomorrow.org, 2013 (www.tomorrow.org/speakup/2013_OnlineLearningReport.html), p. 7.

22. Project Tomorrow, "From Chalkboard to Tablets: The Emergence of the K–12 Digital Learner," p. 10.

23. Ibid., p. 12.

24. See the Consortium for School Networking website at https://consortiumforschoolnetworking167.eduvision.tv/Default.aspx?q=IsahX h4JBPQIfCcXr07jCyH71H4JEBd8.

25. Keith Krueger quoted in Wireless Reach EdTech Conference Proceeding, "Teachers' Views on Mobile Learning in the Classroom," Washington, October 11–12, 2013, p. 40.

26. Project Tomorrow, "The Onslow County 1-to-1 Math Initiative: Leveraging Mobile Devices to Transform Teaching and Learning in High School Math Classes," October 2012 (www.tomorrow.org/docs/Onslow%20External%20Evaluation%20Report%20Oct%202012.pdf), and Qualcomm Wireless Reach, "24/7 Wireless Collaboration and

Teaching Enhances Student Engagement and Math Development" (San Diego, July 2012).

27. Esther Wojcicki, "Designing K–12 Education for the Innovation Economy," paper presented at the i4j Summit, Menlo Park, Calif., March 2013, pp. 5–6.

28. Market Watch, "Harvard Prof Chris Dede Encourages Teachers to Embrace Mobile Online Learning," MarketWatch.com, November 16, 2011.

29. Edward Wyatt, "F.C.C. Backs Plan to Update a Fund That Helps Connect Schools to the Internet," *New York Times,* July 20, 2013, p. B3.

30. Julie Evans, "Making Learning Mobile: Stone Middle School Preliminary Data Findings," unpublished paper, June 2013.

31. Byron August, Paul Kihn, and Matt Miller, "Closing the Talent Gap: Attracting and Retaining Top-Third Graduates to Careers in Teaching" (New York: McKinsey & Co., September 2010) (http://mckinseyonsociety.com/downloads/reports/Education/Closing_the_talent_gap.pdfp), p. 5.

32. West, *Digital Schools.*

Chapter 6

1. Hao Wang and Jing Liu, "Mobile Phone Based Health Care Technology," *Recent Patents in Biomedical Engineering* 2 (2009): 15–21.

2. Portions of this chapter are drawn from Darrell West, "How Mobile Devices Are Transforming Healthcare," Brookings Policy Report (Brookings Institution, May 22, 2012).

3. World Health Organization, "mHealth: New Horizons for Health through Mobile Technologies," Global Observatory for eHealth Series, Geneva, Switzerland, vol. 3, 2011.

4. Robert Litan, "Vital Signs via Broadband: Remote Monitoring Technologies Transmit Savings," Better Health Care Together Coalition, October 24, 2008 (www.corp.att.com/healthcare/docs/litan.pdf), p. 1.

5. Telephone interview with Amy Waterman, Erica Winston, and Blake Thai, March 13, 2012.

6. Qualcomm Wireless Reach, "Wireless Heart Health: Using 3G to Assist Underserved Patients with Cardiovascular Disease" (San Diego, December 2011) (www.qualcomm.com/#/news/releases/2011/09/07/qualcomm-and-life-care-networks-launch-3g-mobile-health-project-help).

7. Edward Boyer, Rich Fletcher, Richard Fay, David Smelson, Douglas Ziedonis, and Rosalind Picard, "Preliminary Efforts Directed toward the Detection of Craving of Illicit Substances: The iHeal Project," *Journal of Medical Toxicology* 8 (February 2012): 5–9.

8. Lorien Abroms, Nalini Padmanabhan, Lalida Thaweethai, and Todd Phillips, "iPhone Apps for Smoking Cessation," *American Journal of Preventive Medicine* 40 (2011): 279–85.

9. Laura Green, "Boca Cardiologist Develops Healthy Choices Smartphone App," *Palm Beach Post*, March 16, 2012.

10. James G. Kahn, Joshua Yang, and James S. Kahn, "'Mobile' Health Needs and Opportunities in Developing Countries," *Health Affairs*, February 2010, p. 258.

11. Amy Cueva, "Mobile Technology for Healthcare: Just What the Doctor Ordered?," *UX Magazine*, September 14, 2010.

12. Mirela Prgomet, Andrew Georgiou, and Johanna Westbrook, "The Impact of Mobile Handheld Technology on Hospital Physicians' Work Practices and Patient Care," *Journal of the American Medical Informatics Association* 16, no. 6 (November–December 2009): 792–801.

13. Kenya Beard, Sue Greenfield, Elsa-Sophia Morote, and Richard Walter, "Mobile Technology: Lessons Learned Along the Way," *Nurse Educator* 36 (May–June 2011): 105.

14. Santosh Krishna, Suzanne Boren, and Andrew Balas, "Healthcare via Cell Phones," *Telemedicine and e-Health*, April 2009, p. 231.

15. Robert Hurling, Michael Catt, Marco De Boni, Bruce Fairley, Tina Hurst, Peter Murray, Alannah Richardson, and Jaspreet Sodhi, "Using Internet and Mobile Phone Technology to Deliver an Automated Physical Activity Program," *Journal of Medical Internet Research* 2 (April–June 2007) (www.jmir.org/2007/2/e7/).

16. Catharine Paddock, "iPads in Health and Medicine," *Medical News Today*, March 14, 2012.

17. Meredith Cohn, "Hopkins Researchers Aim to Uncover Which Mobile Health Applications Work," *Baltimore Sun*, March 14, 2012.

18. Jose Marquez, "Will mHealth Revolutionize Healthcare?," *Huffington Post*, March 7, 2012.

19. Telenor Group, "mHealth Partnership Supports Mother-Infant Health," May 31, 2011.

20. Christen Brownlee, "mHealth—Can You Hear Me Now?," *Magazine of the Johns Hopkins University Bloomberg School of Public Health*, 2012 (magazine.jhsph.edu/2012/technology/features/mHealth/page_1/).

21. Ibid.

22. World Health Organization, *Adherence to Long-Term Therapies: Evidence for Action,* WHO.int, 2003 (www.who.int/chp/knowledge/publications/adherence_full_report.pdf), pp. 7–10.

23. Boston Consulting Group and Telenor Group, "Socio-Economic Impact of mHealth," February 28, 2012.

24. Qualcomm Wireless Reach Initiative, "Japan: Wireless_Health_Care@Home, Enabling 3G Health Care Access for Rural Communities," June 2011.

25. Rediff.com, "Why Healthcare Sector Is Upbeat about Telemedicine," March 14, 2012.

26. SingHealth, "Health Buddy—Health Mobile Application," March 20, 2012.

27. Chris Sweeney, "How Text Messages Could Change Global Healthcare," *Popular Mechanics,* October 24, 2011.

28. Nuwan Waidyanatha, Arur Dubrawski, Ganesan M., and Gordon Gow, "Affordable System for Rapid Detection and Mitigation of Emerging Diseases," *International Journal of E-Health and Medical Communications* 2 (January–March 2011): 73–90.

29. Boston Consulting Group and Telenor Group, "Socio-Economic Impact of mHealth."

30. Darrell West and Edward Alan Miller, *Digital Medicine: Health Care in the Internet Era* (Brookings Institution Press, 2009), p. 4.

31. James Katz and Ronald Rice, "Public Views of Mobile Medical Devices and Services," *International Journal of Medical Informatics* 78 (2009): 104–14.

32. Bipartisan Policy Center Task Force on Delivery System Reform and Health IT, "Transforming Health Care: The Role of Health IT," January 2012.

33. Brett Norman, "FDA Tangles With Wireless Medical-App Makers," *Politico,* April 17, 2012, p. 26.

34. Brian Dolan, "FDA Mulls the Role of Screening Apps, Devices," *Mobile Health News,* March 20, 2012.

35. Warren Kaplan, "Can the Ubiquitous Power of Mobile Phones Be Used to Improve Health Outcomes in Developing Countries?," *Globalization and Health,* May 23, 2006.

36. Patricia Mechael, Hima Batavia, Nadi Kaonga, Sarah Searle, Ada Kwan, Adina Goldberger, Lin Fu, and James Ossman, "Barriers and Gaps Affecting mHealth in Low and Middle Income Countries" (New

York: Columbia University Center for Global Health and Economic Development, May 2010).

37. PricewaterhouseCoopers report quoted in Gulveen Aulakh, "Mobile Health to Be Rs 3,000 Core Market in India by 2017," *Economic Times,* March 2, 2012.

38. Roger Entner, "The Wireless Industry: The Essential Engine of U.S. Economic Growth," ReconAnalytics.com, May 2012, pp. 30–33.

39. Harald Gruber and Pantelis Koutroumpis, "Mobile Telecommunications and the Impact on Economic Development," Panel Draft for 52nd Economic Policy Panel, hosted by the Einaudi Institute for Economics and Finance, Rome, October 22–23, 2010, in cooperation with the Centre for Economic Policy Research, London (http://dev3.cepr.org/meets/wkcn/9/979/papers/Gruber_Koutroumpis.pdf).

40. Leonard Waverman, Meloria Meschi, and Melvyn Fuss, "The Impact of Telecoms on Economic Growth in Developing Countries," Vodafone Policy Papers, Vodafone.com, 2005.

Chapter 7

1. Portions of this chapter are drawn from Darrell West, "Improving Health Care through Mobile Medical Devices and Sensors," Brookings Policy Report (Brookings Institution, October 22, 2013).

2. Errol Ozdalga, Ark Ozdalga, and Neera Ahuja, "The Smartphone in Medicine," *Journal of Medical Internet Research* 14 (September 27, 2012).

3. Timothy Aungst, "Study Suggests Researchers Should Use Social Media 'App' Websites to Engage Patients in Disease Surveillance," iMedicalApps.com, May 28, 2013.

4. Brian Dolan, "Why the Qualcomm Life 2net Launch Matters," *Mobile Health News,* December 8, 2011.

5. Duck Lee, Jaesoon Choi, Ahmed Rabbi, and Reza Fazel-Rezai, "Development of a Mobile Phone Based e-Health Monitoring Application," *International Journal of Advanced Computer Science and Applications* 3 (2012): 38–43.

6. Brian Dolan, "Asthmapolis Secures FDA Clearance for Inhaler Sensor," MobileHealthNews.com, July 11, 2012.

7. Nancy Hudecek, "Patient Safety and Physician Satisfaction," AirStrip White Paper Series, AirStrip.com, 2013, p. 4.

8. Suneet Chauhan, Han-Yang Chen, Cande Ananth, Anthony Vintzileos, and Alfred Abuhamad, "Electronic Fetal Heart Rate Monitoring Greatly Reduces Infant Mortality," *Science Daily*, February 2011. Paper presented at the annual meeting of the Society for Maternal-Fetal Medicine, San Francisco, 2011.

9. Oguz Karan, Canan Bayraktar, Haluk Gumuskaya, and Bekir Karlik, "Diagnosing Diabetes Using Neural Networks on Small Mobile Devices," *Expert Systems with Applications* 39 (2012): 54.

10. David Hroncheck, "Zephyr's HxM Bluetooth Heart Rate Monitor," *Running Digital,* January 6, 2010.

11. U.S. Agency for International Development, "mPowering Frontline Health Workers," June 14, 2012 (www.usaid.gov/news-information/press-releases/mpowering-frontline-health-workers).

12. Qualcomm, "3G Wireless Technology Provides Clinical Information to Public Health Care Workers through Mobile Health Information System Project," press release, Qualcomm.com, November 10, 2010 (www.qualcomm.com/media/releases/2010/11/10/3g-wireless-technology-provides-clinical-information-public-health-care).

13. See a description of this program at www.qualcomm.com/media/documents/wireless-reach-case-study-japan-wireless-health-care-english.

14. Jon Gertner, "Meet the Tech Duo That's Revitalizing the Medical Device Industry," *Fast Company,* April 15, 2013 (www.fastcompany.com/3007845/meet-tech-duo-thats-revitalizing-medical-device-industry).

15. U.S. Food and Drug Administration, "Classify Your Medical Device," December 3, 2012 (www.fda.gov/MedicalDevices/DeviceRegulationandGuidance/Overview/ClassifyYourDevice/default.htm).

16. Daniel Kramer, Shuai Xu, and Aaron Kesselheim, "Regulation of Medical Devices in the United States and European Union," *New England Journal of Medicine* 366 (March 1, 2012): 848.

17. Ibid., p. 849.

18. Ibid., p. 850.

19. Frederic Resnic and Sharon-Lise Normand, "Postmarketing Surveillance of Medical Devices: Filling in the Gaps," *New England Journal of Medicine* 366 (March 8, 2012): 875.

20. U.S. Food and Drug Administration, "Medical Device User Fee Amendments of 2012," August 3, 2012.

21. The press release on the final MMA guidance document may be found at www.fda.gov/NewsEvents/Newsroom/PressAnnouncements/ucm369431.htm, while the final guidance itself may be found at www.

fda.gov/downloads/MedicalDevices/DeviceRegulationandGuidance/ GuidanceDocuments/UCM263366.pdf.

22. U.S. Food and Drug Administration, "Mobile Medical Applications: Guidance for Industry and Food and Drug Administration Staff," September 23, 2013, pp. 20–22.

23. Ibid., pp. 26–28.

24. U.S. Food and Drug Administration, "FDA Issues Final Guidance on Mobile Medical Apps," press release (GPO, September 23, 2013).

25. U.S. Food and Drug Administration, "Mobile Medical Applications," pp. 23–25.

Chapter 8

1. Portions of this chapter are drawn from Darrell West, "Facebook, iPhones: How Evolving Mobile Technology Shapes Campaigns," Brookings Policy Report (Brookings Institution, February 14, 2012).

2. Charlie Warzel, "Romney Goes Mobile, Uses Jumptap to Target Voters in Iowa & New Hampshire," BostInno.com, January 10, 2012.

3. Lauren Johnson, "Mobile Will Play Pivotal Role in 2012 Political Campaigns," *Mobile Marketer*, January 10, 2012.

4. Lauren DeLisa Coleman, "2012 Candidates Playing Catchup with Mobile Tech-Savvy Voters," *Huffington Post*, September 12, 2011, and John Chang, "Mobile Technology and Its Impact on Political Campaigns," *PC Tech Mojo*, January 13, 2012.

5. Lee Rainie and Aaron Smith, "Politics Goes Mobile," Pew Internet & American Life Project (Washington: Pew Research Center, December 23, 2010).

6. Shannon Young, "Candidates Use Latest Technology to Campaign," *Nashua Telegraph*, January 9, 2012.

7. Marshall Kirkpatrick, "New Mobile App Makes Billboards Talk about European Politics," *ReadWriteWeb*, August 5, 2011.

8. King's College London, "Informatics Students Win 'App' Design Competition and Meet Prime Minister" (London, January 31, 2012).

9. Chris Good, "Walking Edge: Canvassing With GPS," *The Atlantic*, January 2010.

10. Andrew Romano, "Yes We Can (Can't We?): Team Obama Has Quietly Built a Juggernaut Reelection Machine in Chicago," *The Daily Beast*, January 2, 2012.

11. Ibid.

12. Nancy Scola, "Doing Digital for Romney: An Interview with Zac Moffatt," *The Atlantic,* January 2012.

13. United Nations Development Programme, "Haiti Elections: Cell Phones and Internet to Facilitate Voter Turnout," UNDP press release, March 18, 2011 (www.undp.org/content/undp/en/home/press-center/articles/2011/03/18/haiti-elections-cell-phones-and-internet-to-facilitate-voter-turnout).

14. Miroslav Cepicky, "Mobile Voting Would Attract an Additional Half Million Voters," Vodafone press release, May 27, 2010.

15. Emily Schultheis, "Political Ads Go Mobile," *Politico,* November 28, 2011.

16. Ibid.

17. "GOP Presidential Hopefuls Experiment with Mobile Ads in Early Contests," *Advertising Age,* January 23, 2012 (http://adage.com/article/campaign-trail/gop-presidential-hopefuls-experiment-mobile-ads/232284/).

18. Ibid.

19. Cami Zimmer quoted in Brenda Krueger Huffman, "Mobile Apps—a Must for 2012 Political Campaigns," *Axcess News,* January 20, 2012.

20. Dan Butcher, "Mobile Ad Campaigns 5 Times More Effective Than Online," *Mobile Marketer,* February 5, 2010.

21. Aaron Smith, "Real Time Charitable Giving," Pew Internet & American Life Project (Washington: Pew Research Center, January 12, 2012).

22. Annie Linskey, "Rule Would Allow Campaign Donations by Text Message," *Baltimore Sun,* December 1, 2011.

23. Smith, "Real Time Charitable Giving."

24. Young, "Candidates Use Latest Technology to Campaign."

25. Timothy Fleming, "FEC Shoots Down Proposal for Texted Campaign Contributions," ABC News, December 12, 2010.

26. Scott Goodstein, "Mobile vs. FEC," *Huffington Post,* December 22, 2010.

27. Byron Tau, "Obama Campaign Rolls Out Square Mobile Fundraising Platform," *Politico,* January 30, 2012.

28. Nick Judd, "Obama Campaign Using Square to Collect Campaign Donations," *TechPresident,* January 30, 2012.

29. Linskey, "Rule Would Allow Campaign Donations by Text Message."

30. Michelle Quinn, "California OKs Donations via Text," *Politico,* October 13, 2011.

31. "Political Party Donations Go Mobile," *BizCommunity,* February 2, 2009.

32. Tony Dennis, "Grapple Produces Election Mobile App for Greens," *GreensGoMo News,* April 27, 2011.

33. Information provided by Sarah Wheaton, editor of the New York Times Election 2012 app, February 7, 2012.

34. Charles Gibson, "How Wireless Technology Supports Democracy," PowerPoint presentation, University of California at San Diego, 2011.

35. Ian Schuler, "SMS as a Tool in Election Observation," *Innovations,* Spring 2008.

36. Ibid.

Chapter 9

1. Intergovernmental Panel on Climate Change, *Managing the Risks of Extreme Events and Disasters to Advance Climate Change Adaptation* (Cambridge University Press, 2012).

2. Ibid.

3. Ibid.

4. Darrell West and Marion Orr, "Race, Gender, and Communications in Natural Disasters," *Policy Studies Journal* 35, no. 4 (November 2007): 569–86.

5. Jason Om, "Mobile Invention Could Be Desert Lifeline," *ABC Science,* July 12, 2010 (www.abc.net.au/science/articles/2010/07/12/2951206.htm#.Ucxwlz7wKrh).

6. Nichole Laporte, "From One Tragedy, Tools to Fight the Next," *New York Times,* January 12, 2012 (www.nytimes.com/2012/01/22/business/inventions-offer-tools-to-endure-future-disasters.html?_r=0).

7. "Public-Safety Mobile App Hackathon Announces Winners," Radio Resource Media Group news brief, May 24, 2013 (www.rrmedia group.com/newsArticle.cfm?news_id=9566).

8. Michael Agger, "Japan's Earthquake and Tsunami: How Google Responds to Crises, Plus Good Emergency Apps for Your Phone," Slate.com, March 11, 2011 (www.slate.com/articles/technology/the_browser/2011/03/earthquake_japan_help.html).

9. Qualcomm Wireless Reach, "Let's Get Ready! Mobile Safety Project," Qualcomm, June 2013 (www.qualcomm.com/media/documents/files/wireless-reach-case-study-china-let-s-get-ready-english-.pdf).

10. "Disaster-Related Mobile Apps from Japan," *Emergency Journalism,* November 26, 2012.

11. Marc Saltzman, "Japan Quake Popularizes Disaster Apps," *USA Today,* March 20, 2011.

12. David Burns, "Emergency Management Apps: A Primer," *Campus Safety Magazine,* November 29, 2011.

13. U.S. Department of Health and Human Services, "Disaster Information Management Research Center: Disaster Apps and Mobile Optimized Web Pages" (GPO, July 12, 2012).

14. Nita Bhalla, "Disaster-Prone Bangladesh Trials Early Warning Cell Phone Alerts," Thomson Reuters Foundation, June 23, 2009 (http://in.reuters.com/article/2009/06/23/idINIndia-40548720090623).

15. "Brigade to Lead the Way on Emergency Tweets," press release, *London Fire Brigade,* December 8, 2012 (www.london-fire.gov.uk/news/LatestNewsReleases_1812201220.asp#.UcxyuD7wKrg).

16. See table 1 in World Economic Forum Report (http://reports.weforum.org/global-information-technology-2012/).

17. "The Transformational Use of Information and Communication Technologies in Africa," World Bank Group (New York, December 10, 2012) (http://siteresources.worldbank.org/EXTINFORMATIONANDCOMMUNICATIONANDTECHNOLOGIES/Resources/282822-1346223280837/Summary.pdf).

18. "Flood, Famine and Mobile Phones," *The Economist,* July 26, 2007.

19. Timothy Spence, "African Herdsmen Use Mobile Phones for Drought Alerts," *EurActiv,* April 15, 2013.

20. "European Commission Fact Sheet: Horn of Africa Drought: Covering Kenya, Ethiopia, Somalia and Djibouti," European Commission Humanitarian Aid and Civil Protection, November 11, 2011 (http://ec.europa.eu/echo/files/aid/countries/hoa_drought_factsheet.pdf).

21. Nancy McNally, "Cell Phones Revolutionizing Kenya's Livestock Sector," Food and Agricultural Organization of the United Nations, March 1, 2012, web June 25, 2013 (www.fao.org/news/story/en/item/170807/icode/).

22. "Flood, Famine and Mobile Phones," *The Economist.*

23. "Report of the Campus Safety Task Force Presented to Attorney General Roy Cooper," Attorney General Roy Cooper Campus Safety Task Force, January 2008 (http://vptm.ehps.ncsu.edu/files/2213/2750/6549/CampusSafetyReport_2008.pdf).

24. "North Carolina Department of Justice: Campus Safety News Report" (www.ncdoj.gov/Top-Issues/School-Safety/Campus-Safety.aspx).

25. Zoe Fox, "Virginia Tech Shooting Survivor Launches Campus Safety App," *Mashable,* April 20, 2013.

26. Maroon Editorial Board, "Safety in Knowledge," *Chicago Maroon* (University of Chicago), May 14, 2013.

27. "UNCC Officers Test New App during Training," *Campus Safety Magazine,* July 31, 2012.

28. "Wash. District Implements Mobile Surveillance Application," *Campus Safety Magazine,* August 12, 2012.

29. Amy Canfield, "Campus Mass Notification Evolves," *Security Director News,* January 11, 2013.

30. Ibid.

31. Whit Richardson, "Schools Seek 'Easy Button' Approach to Mass Notification Systems," *Security Systems News,* February 21, 2012.

32. Canfield, "Campus Mass Notification Evolves."

33. Ibid.

34. National Commission on Terrorist Attacks Upon the United States, *The 9/11 Commission Report* (Washington: GPO, 2004).

35. "Issue Brief: 2007 State Homeland Security Directors Survey," National Governors Association Center for Best Practices, December 18, 2007 (www.nga.org/files/live/sites/NGA/files/pdf/0712HOMELAND SURVEY.PDF).

36. "Report of the Campus Safety Task Force Presented to Attorney General Roy Cooper."

37. Michael Cooney, "IBM Says Software Helps Predict Natural Disasters," NetworkWorld.com, October 23, 2010 (www.networkworld.com/community/blog/ibm-says-software-helps-predict-natural-disasters).

38. Tom Pica, "The New Jersey Shore Days after Hurricane Sandy," Verizon Wireless News Center, November 20, 2012.

39. PBS, "Mobile Emergency Alert System Promises Flexible New Option for Emergency Managers and First Responders," LG Harris and PBS, August 20, 2012 (http://mobileeas.org/wp-content/uploads/2013/06/MEAS-at-APCO-Minneapolis-FINAL-copy.pdf).

40. "Raytheon Releases One Force Mobile Collaboration App for First Responders," *PR Newswire,* February 11, 2013.

41. "PremierOne Mobile," *Motorola Solution* (www.motorola solutions.com/US-EN/Business+Solutions/Industry+Solutions/Government/Law+Enforcement/PremierOne_Mobile_US-EN).

42. KathyMarks, "Emergency-Response Mobile Applications," *Law and Order: The Magazine for Police Management,* April 2013.

43. National Crimes Records Bureau, Ministry of Home Affairs, "Crime in India 2012" (New Delhi: Ministry of Home Affairs, 2012) (http://ncrb.gov.in/).

44. "Nasscom Felicitates Women Safety Mobile App Makers," *Economic Times,* June 19, 2013.

45. Qualcomm Wireless Reach report, "Strengthening Crime Mapping through Telecommunications Technology in El Salvador," Qualcomm. com (www.qualcomm.com/media/documents/files/wireless-reach-case-study-el-salvador-wireless-security-english-.pdf).

46. Niltin Puri, "Fight Back with Mobile Safety App Fightback," *ZD Net,* March 8, 2013.

47. Clarice Africa, "Taiwan launches Mobile App for Public Safety," *Mobile Government,* August 27, 2012 (www.futuregov.asia/articles/2012/aug/27/taiwan-launches-mobile-app-public-safety/).

48. Annie Kelly, "Kenya Turns to Mobile App to Stop motorbike mayhem on the roads," *The Guardian,* May 13, 2013.

49. Federal Communications Commission, "U.S. and Canada Reach Agreement on Border Spectrum Sharing Arrangements," May 14, 2013 (www.fcc.gov/document/us-canada-reach-agreement-border-spectrum-sharing-arrangements).

50. Phil Kurz, "U.S.-Canada Reach Deal on Mobile, Public Safety Spectrum Sharing," *Broadcast Engineering,* May 16, 2013.

51. Ida Torres, "Japan to Learn from Taiwan in Disaster-Proofing Telecoms," *Japan Daily Press,* February 13, 2012.

52. Ibid.

53. Sung Szjie, "How Can We Use Mobile Apps for Disaster Communications in Taiwan: Problems and Possible Practice," *EconStor,* June 2011 (www.econstor.eu/dspace/bitstream/10419/52323/1/67297973X.pdf).

54. Ibid.

55. Ibid.

56. Axel Bruns and Yixian Eugene Liang, "Tools and Methods for Capturing Twitter Data during Natural Disasters," *First Monday,* April 17, 2012.

57. Szjie, "How Can We Use Mobile Apps for Disaster Communications in Taiwan."

58. Claudia Pagliari and Francesco Pinciroli, "Mobile Emergency, and Emergency Support System for Hospitals in Mobile Devices: Pilot

Study," *JMIR* (*Journal of Medical Internet Research*), May 23, 2013 (www.ncbi.nlm.nih.gov/pubmed/23702566).

59. Alvaro Monares and others, "Mobile Computing in Urban Emergency Situations: Improving the Support to Firefighters in the Field," *Expert Systems with Applications* 38, no. 2 (February 2011): 1255–67.

60. Ibid.

61. Emmanuel Lallana, "eGovernment for Development," Institute for Development Policy and Management, October 19, 2008 (www.egov4dev.org/mgovernment/applications/index.shtml).

62. Ibid.

63. Britta Glennon, "The Role of Technology in Crisis Management and How It Could Be Done Better," *Chicago Policy Review*, May 7, 2013 (http://chicagopolicyreview.org/2013/05/07/the-role-of-technology-in-crisis-management-and-how-it-could-be-done-better/).

64. Om, "Mobile Invention Could Be Desert Lifeline."

65. Shawn Knight, "Hurricane Sandy Damages Crucial Wireless and Internet Infrastructure," *Techspot,* October 31, 2013 (www.techspot.com/news/50668-hurricane-sandy-damages-crucial-wireless-and-internet-infrastructure.html).

66. Matt Wood and Alexandra Martines, "Why Did Wireless Networks Fail after Hurricane Sandy?," Freepress.net, November 21, 2012 (www.savetheinternet.com/blog/2012/11/21/why-did-wireless-networks-fail-after-hurricane-sandy).

67. Gerry Smith, "Oklahoma City Area Hit by Phone, Internet Outages after Tornado," *Huff Post Tech,* May 21, 2013 (www.huffingtonpost.com/2013/05/21/oklahoma-city-phone-internet-outages_n_3312790.html).

68. "Before the Federal Communications Commission, Washington, D.C. 20554: Comments of the CTIA—The Wireless Association," PS docket no. 11-60, PS docket no. 10-92, EB docket no. 06-119, July 7, 2011, accessed June 25, 2013 (http://files.ctia.org/pdf/filings/110707_-_FILED_CTIA_Network_Reliability_Comments.pdf).

69. International Telecommunication Union, "ITU Joins International Effort to Assist Haiti," press release (www.itu.int/newsroom/press_releases/2010/02.html).

70. Greg Slabodkin, "National Wireless Communications System for First Responders Years Away," *Fierce Mobile Healthcare,* December 10, 2012 (www.fiercemobilehealthcare.com/story/national-wireless-communications-system-first-responders-years-away/2012-12-10).

71. U.S. Department of Commerce, "Fact Sheet: First Responder Network Authority (FirstNet)," August 20, 2012 (www.commerce.gov/news/fact-sheets/2012/08/20/fact-sheet-first-responder-network-authority-firstnet).

Chapter 10

1. Kevin Ashton, "That 'Internet of Things' Thing: In the Real World Things Matter More Than Ideas," *RFID Journal,* June 22, 2009.

2. Nam Pham, "The Economic Benefits of Commercial GPS Use in the U.S. and the Costs of Potential Disruption" (Washington: NDP Consulting, June 2011).

3. Maravedis Rethink, "Mobile Operators Strategic Analysis Quarterly Report," November 2012.

4. A. T. Kearney, "GSMA The Mobile Economy" (London, 2013).

5. Darrell West, "How Mobile Technology Is Driving Global Entrepreneurship," Brookings Policy Report (Brookings Institution, October 23, 2012).

6. Maja Andjelkovic and Saori Imaizumi, "Maximizing Mobile: Mobile Entrepreneurship and Employment," World Bank Information and Communications for Development (New York, 2012), p. 76. See http://siteresources.worldbank.org/EXTINFORMATIONAND COMMUNICATIONANDTECHNOLOGIES/Resources/IC4D-2012-Report.pdf.

7. Harald Gruber and Pantelis Koutroumpis, "Mobile Telecommunications and the Impact on Economic Development," *Economic Policy* 67 (July 2011): 387–426.

8. Ibid.

9. LECG, "3G Mobile Networks in Emerging Markets: The Importance of Timely Investment and Adoption," January 26, 2009 (https://observatorio.iti.upv.es/media/managed_files/2009/03/09/Report_GSMA_LECG_Feb09.pdf). Quotation from link at http://www3.weforum.org/docs/GITR/2012/GITR_Chapter1.5_2012.pdf.

10. Analysis Mason, "Assessment of Economic Impact of Wireless Broadband in India," December 2010 (www.gsma.com/spectrum/wp-content/uploads/2012/03/amindiaexecsummaryfinal.pdf).

11. GSM Association, "The European Mobile Manifesto: How Mobile Will Help Achieve Key European Union Objectives," October 2009 (www.itu.int/ITU-D/partners/GILF/2009/docs/GSMA.pdf).

12. Deloitte, "The Impact of 4G Technology on Commercial Interactions, Economic Growth, and U.S. Competitiveness" (Chicago, August 2011) (www.deloitte.com/assets/DcomUnitedStates/Local%20Assets/Documents/TMT_us_tmt/us_tmt_impactof4g_081911.pdf).

13. GSMA and the Boston Consulting Group, "The Economic Benefits of Early Harmonization of the Digital Dividend Spectrum & the Cost of Fragmentation in Asia Pacific (www.gsma.com/spectrum/wp-content/uploads/2012/07/277967-01-Asia-Pacific-FINAL-vf11.pdf).

14. Deloitte, "Value of Connectivity: Economic and Social Benefits of Expanding Internet Access," February 2014 (https://fbcdn-dragon-a.akamaihd.net/hphotos-ak-ash3/t39.2365/851546_1398036020459876_1878998841_n.pdf).

15. See the Technology for Transparency Network at http://transparency.globalvoicesonline.org/about.

16. Eric Mack, "Facing a Mobile Bandwidth Drought," *PC World* 29 (November 2011): 13–14.

17. Sean Xin Xu, Yan Xu, and Xiaona Zheng, "Communication Platforms in Electronic Commerce: A Three-Dimension Analysis," *Info: The Journal of Policy, Regulation and Strategy for Telecommunications, Information and Media* 10 (2008): 47–56.

INDEX